ENVIRONMENTAL TOXICOLOGY AND PHARMACOLOGY OF HUMAN DEVELOPMENT

ENVIRONMENTAL TOXICOLOGY AND PHARMACOLOGY OF HUMAN DEVELOPMENT

Edited by

Sam Kacew, Ph.D.
Department of Pharmacology
University of Ottawa
Ottawa, Ontario, Canada

George H. Lambert, M.D.
Department of Pediatrics
Division of Pediatric Pharmacology and Toxicology
University of Medicine and Dentistry of New Jersey
Piscataway, New Jersey

Taylor & Francis
Publishers since 1798

USA	Publishing Office:	Taylor & Francis 1101 Vermont Avenue, NW, Suite 200 Washington, DC 20005-3521 Tel: (202) 289-2174 Fax: (202) 289-3665
	Distribution Center:	Taylor & Francis 1900 Frost Road, Suite 101 Bristol, PA 19007-1598 Tel: (215) 785-5800 Fax: (215) 785-5515
UK		Taylor & Francis Ltd. 1 Gunpowder Square London EC4A 3DE Tel: 171 583 0490 Fax: 171 583 0581

ENVIRONMENTAL TOXICOLOGY AND PHARMACOLOGY OF HUMAN DEVELOPMENT

1 2 3 4 5 6 7 8 9 0   B R B R   9 8 7

This book was set in Times Roman. The editors were Elizabeth Dugger and Catherine Simon. Cover design by Michelle Fleitz.

A CIP catalog record for this book is available from the British Library.

∞ The paper in this publication meets the requirements of the ANSI Standard Z39.48-1984 (Permanence of Paper)

Library of Congress Cataloging-in-Publication Data

Environmental toxicology and pharmacology of human development/
    edited by Sam Kacew, George H. Lambert.
        p. cm.
    Includes bibliographical references and index.

        1. Perinatal pharmacology.   2. Fetus—Effect of chemicals on.
3. Breast milk—Contamination.   4. Pediatric toxicology.   I. Kacew,
Sam.   II. Lambert, George H.
    [DNLM: 1. Fetus—drug effects.   2. Infant, Newborn.
3. Environmental Exposure—in pregnancy.   4. Environmental Exposure—
in infancy & childhood.   5. Drugs—adverse effects.   WQ 210 E61
1997]
RG627.6.D79E58   1997
618.3'2—DC20
DNLM/DLC
for Library of Congress   96-43034
                                                                                                            CIP

ISBN 1-56032-448-1

Contents

Contributors

GEORGIANNE ARNOLD, M.D.
Division of Pediatric Genetics
University of Rochester Medical Center
601 Elmwood Avenue, Box 777
Rochester, NY 14642

CHESTON M. BERLIN, Jr., M.D.
Department of Pediatrics
Milton S. Hershey Medical Center
Pennsylvania State University College
   of Medicine
P.O. Box 850
Hershey, PA 17033-0850

SAM KACEW, Ph.D.
Department of Pharmacology
University of Ottawa
Ottawa, Ontario, Canada K1H 8M5

MICHAEL D. KOGAN, Ph.D.
U.S. Center for Disease Control
   and Prevention
National Center for Health Statistics
Presidential Building
6525 Belcrest Road
Hyattsville, MD 20782

JOAN CREECH KRAFT, R.N., Ph.D.
Department of Pharmacology
University of Washington
Seattle, WA 98195

JEREMIAH J. LEVINE, M.D.
Division of Gastroenterology
   and Nutrition
Schneider Children's Hospital, Room 229
New Hyde Park, NY 11040

JUDITH S. SCHREIBER, Ph.D.
Bureau of Toxic Substance Assessment
New York State Department of Health
2 University Place, Room 240
Albany, NY 12203

BENNETT A. SHAYWITZ, M.D.
Department of Pediatrics
Neurology and Child Study Center
School of Medicine
Yale University
3089 IMP
P.O. Box 3333
New Haven, CT 06510

CHARLES V. SMITH, Ph.D.
Department of Pediatrics
Baylor College of Medicine
One Baylor Plaza
Houston, TX 77030

MICHAEL B. H. SMITH, M.B., B.Ch.,
   F.R.C.P.C.
Division of Pediatric Medicine
Dalhousie University
IWK-Grace Health Centre
P.O. Box 3070
5850 University Avenue
Halifax, Nova Scotia, Canada B3J 3G9

CALVIN C. WILLHITE, Ph.D.
Department of Toxic Substances Control
State of California
700 Heinz Street, Suite 200
Berkeley, CA 94710

Preface

Society derives great benefit from the use of chemicals in agriculture, energy production, transportation, pharmaceuticals, and other products that improve the quality of life. However, the extensive use of chemicals and pharmaceutical products can result in inadvertent exposure of the fetus and the neonate. Clearly today's society is highly dependent on a vast array of chemicals and chemical processes to maintain or improve the current standard of living. It is essential, however, to minimize potential adverse effects to the fetus and the developing newborn.

The fetus and the neonate of both animals and humans are uniquely different from the respective adults in their responses to many chemicals, including innumerable drugs. In Sam Kacew's introductory chapter the basic principles of exposure and factors that impinge upon these parameters are covered from the unborn fetus to the breast-fed infant. The purpose of this chapter is to review current knowledge of the pharmacokinetic characteristics, metabolism, renal handling, etc. of chemicals and drugs in the fetus, breast-fed infant, and developing newborn. The second chapter, by Joan Creech Kraft and Calvin C. Willhite, deals primarily with the fetus and the manner in which chemicals can affect outcome in the newborn. Excess of the often ingested vitamin A (retinoids) provides an example of abnormal infant development resulting from maternal fetal exposure. Although the Cold War may be over, the environmental contamination battle and adverse consequences on infant growth and development remains in Eastern Europe, as outlined by Charles V. Smith in Chapter 3. The severe pediatric problems currently faced in Eastern Europe, will ultimately affect the West.

Breast-feeding is essential for the normal physical and psychological development of the infant to adulthood. This fact clearly indicates that breast-feeding should always be encouraged. However, breast milk can also serve as a route of exposure to chemicals in general (Chapter 4, by Cheston M. Berlin, Jr., and Sam Kacew); to tetrachloroethylene or dry cleaning fluid in particular (Chapter 5, by Judith S. Schreiber); or to silicone through breast implants (Chapter 7, by

Jeremiah J. Levine). In certain circumstances, one must weigh the benefits of breast milk versus chemical contaminant exposure and adverse effects.

Glue sniffing may be popular in order to reach a "high." However, as Georgianne Arnold points out in Chapter 6, there are dire consequences of maternal solvent abuse on the developing infant. The debate over the issue of aspartame usage and potential toxicity in children is covered by Bennett A. Shaywitz in Chapter 8. Finally, in Chapter 9, Michael B. H. Smith and Michael D. Kogan show that self-medication (over-the-counter, OTC) preparations are not necessarily safe just because these chemicals are obtainable without prescription. Indeed, exposure to OTC preparations is a growing problem in our society due to easy access.

Although this monograph was written primarily for the benefit of pharmacologists, toxicologists, and other health scientists, members of the medical profession such as obstetricians, neonatologists, and pediatricians will find chapters of specific interest to their areas of expertise. A wide range of topics is covered, allowing the specialist to gain a broader perspective of fetal and neonatal pharmacology and toxicology.

Sam Kacew
George H. Lambert

Chapter 1

General Principles in Pediatric Pharmacology and Toxicology

Sam Kacew

The functional development of growth of an embryo from conception to birth and postnatally does not occur in a vacuum but is continuously affected by the environment. The study of abnormal responsiveness of newborns directly to therapeutic agents, recreational chemicals, drugs of abuse, or inadvertent exposure to environmental chemicals is termed neonatal toxicology, and there are specific principles that apply uniquely to the newborn. Environmental exposure of newborns arises through the atmosphere from automobile exhaust or from water, food, or ground contamination, be it related to direct spraying of crops by farmers or an industrial accident, such as in Bhopal and Chernobyl, or leaching at a hazardous waste site as occurred at Love Canal. In addition to direct contact of the newborn with chemicals, in utero exposure to drugs or environmental chemicals can also be manifested in altered newborn growth and functional development. Of necessity one must also consider the fact that newborn development can be modified by direct exposure to pharmacological and environmental chemicals through the mother's milk (Kacew, 1993, 1994a). Thus, to be able to adequately evaluate the effects of pharmacological and toxicological agents on the newborn, some consideration must be given to the role of drugs and environmental contaminants during pregnancy as well as after birth (Brent & Beckman, 1994). In our industrialized society it should be noted that the number of women in the work force has increased over the years. Hence included in the definition of environmental exposure is the contact of an individual with chemicals in the workplace, termed occupational exposure. It should be noted that exposure to a chemical during pregnancy may take years before the toxicity becomes overt in the offspring; a notorious example is the use of diethylstilbestrol to prevent miscarriage, which was found 20 years later to

1

result in vaginal cancer in the female offspring and abnormal reproductive tracts in the males. Occupational radiation of adult males can result in the transmission of gene-defective sperm such that cancer develops in the offspring.

Therapeutic agents are designed for the treatment of disorders. A dilemma can arise with the use of drugs for the management of diseases in the mother, as the compounds are beneficial for the mother but can induce severe toxicity in the newborn. Treatment of maternal epilepsy with phenytoin during pregnancy is known to result in congenital anomalies, yet the life of the mother must be considered. Thus, it should be reiterated that therapeutic drugs are, in specific circumstances, administered maternally to treat ailments, yet in the newborn can serve as a source of toxicity. Therapeutic agents are administered directly for the management of newborn disease, yet can in high doses produce adverse toxic reactions. A potentially dangerous but not particularly emphasized problem is the abuse of the "over-the-counter" (OTC) or self-administered drugs as a source of neonatal toxicity (Lock & Kacew, 1988; Kacew, 1994b). On the other hand, recreational or socially used agents including alcohol and tobacco have been recognized as developmental toxicants in the fetus and newborn for many years. In addition to the socially used agents, one should be aware of the high levels of morbidity and mortality among newborn infants passively addicted to drugs of abuse, such as opioids, as a result of maternal drug dependence.

Therapeutic agents form only a small fraction of the total number of chemicals in current use. At present it has been estimated that there are more than 60,000 chemicals in the commercial market; however, less than 20% of these have been evaluated for their toxicological potential in adults. Because it was falsely perceived that an infant was merely a small adult, toxicological data were not obtained for neonates. Hence, substantially less toxicity data are available for newborns exposed to chemicals. With the realization that the newborn responds uniquely to drugs and chemicals, the aim of this chapter is to provide a general background on the developing infant from fetus to neonate and the consequences of toxicant exposure.

CHEMICAL EXPOSURE DURING PREGNANCY

During pregnancy, inadvertent exposure to chemicals will affect two distinct individuals, the mother and fetus (Bottoms et al., 1982). Similarly, the obstetrician in the treatment of disease must consider not only the mother but also the developing fetus. However, little information is available on the influence of chemicals on maternal responses, with consequently even less known about fetal physiology. In reality the effects of chemicals on fetal status can be measured only through indirect techniques in the human. As outlined by Newton (1989), the methodology used to assess fetal health status includes the following:

1 Measurement of various metabolic parameters in maternal serum. This is of particular interest in the measurement of maternal serum α-feto-

protein at 15–20 weeks gestation, an index for a neural tube defect as seen with valproic acid.

2 Aspiration of amniotic fluid (amniocentesis) at mid trimester is used to measure the concentrations of compounds that may predict neonatal distress syndrome, a common index being the lecithin/sphingomyelin ratio. There is also cordocentesis, the aspiration of umbilical cord blood, for the determination of hemolytic disease.

3 Use of ultrasound especially in a high-risk population, such as a patient with hypertension, is very effective in predicting abnormal birth weight.

4 Electronic monitoring of fetal heart rate is indicative of the balance between parasympathetic and sympathetic influences on the heart. A depression in the fetal heart rate can result in hypoxia.

Although these different modes of fetal surveillance are available to the physician for screening infants in the high-risk group, the underlying causative factors involved in fetal injury and toxicity cannot be pinpointed with certainty. Finally it should be emphasized that any diagnostic test can by itself have dire consequences, including maternal morbidity from tocolysis or neonatal morbidity from iatrogenic preterm delivery (Newton, 1989).

A subject that has received little attention but is now recognized as an important factor for fetal growth is the maternal diet and nutritional status. Nutrients essential for fetal growth and development may require an active transport system to be moved from the maternal circulation to the fetal circulation against a concentration gradient (Ginsburg, 1971). However, nutritional deficiency related to chemical or drug ingestion is not uncommon during pregnancy. Basu (1988) reported that chemical-induced vitamin B deficiency in the mother resulted in stillbirth and low-birth-weight infants. A drug-induced deficiency in vitamin A has produced hydrocephalus and ocular defects. Deficiencies in vitamin D caused skeletal abnormalities; in vitamin K produced brain hemorrhage; and in vitamin E caused anencephaly and cleft palate. Nutritional status also affects the process of drug and environmental chemical absorption. In conditions of diets deficient in vitamin D the absorption of lead is increased, resulting in enhanced infant toxicity (Mahaffey, 1980). In the case of drugs, a diet deficient in phosphate can result in renal damage in patients taking high-dose aluminum antacids for treatment of ulcers. A consequence of phosphate deficiency is higher levels of aluminum in both the mother and neonate.

Normally drugs cross the placenta by simple diffusion. A drug-induced destruction of maternal intestinal flora or the induction of vomiting may result in a nutritional imbalance. Consequently, maternal drug levels will rise with higher amounts transported to the fetus. The importance of chemical-induced nutritional deficiency in fetal development is difficult to delineate and more often than not overlooked. However, it is clearly established that in mothers with poor nutritional status, as seen in chronic alcoholism, the morbidity and mortality of neonates are markedly enhanced. During the course of a pregnancy a mother is likely to take a number of drugs for therapeutic reasons. In addition, with many more women in the work force, there is an increased potential of exposure to a variety of chemicals under occupational conditions. Furthermore, a large number of women indulge in a variety of recreational chemicals including cigarettes and alcohol. The consequences attributed to exposure to a pharmaceutical product can be advantageous to the mother; however, in many instances the effects are deleterious to the fetus. Exposure to occupational and/or environmental chemicals is more likely to result in adverse effects than to be beneficial. Thus, it may be stated that the fetus is at some jeopardy as a result of exposure to foreign chemicals. There are a number of factors listed in Table 1-1 that will influence the transfer of chemicals from the maternal circulation to the fetus and will thus have a bearing on any subsequent fetal manifestations.

Table 1-1   Factors Influencing Placental Chemical Transport to the Fetus

Factor	Fetal outcome
Physicochemical properties	The ability of drugs to reach the fetus is dependent on water solubility, lipid solubility, and molecular weight.
Pharmacokinetics	Fetal effects observed are dependent on drug concentration in maternal circulation and in fetal blood supply, on placental and fetal drug metabolism, and on elimination.
Nutrition	Nutritional deficiency in the mother can influence placental transfer of essential nutrients to fetus.
Physiological status	The absence of maternal hormones can alter the ability of fetus to cope with chemicals.
Duration of exposure	A single administration of a high dose of a drug can produce damage to the same extent as chronic, low-dose treatment.
Genetic	Inborn errors of metabolism can predispose a fetus to enhanced toxicity.
Drug interactions	The presence of more than one drug and/or chemical can increase susceptibility of the fetus.
Environment	Various factors in outdoor or indoor environment can modify drug kinetics and thereby affect placental transfer.
Developmental state	Teratogenic agents act selectively on developing cells.
Infection (disease)	Change in maternal body temperature can prolong half-life of drugs, predisposing fetus to enhanced toxicity.
Mechanical	Deformations in structures of uterine tissue affect circulation to placenta and fetus.
Multifactorial (gene–environment interaction)	Various factors in environment in specific, susceptible population can modify drug kinetics and placental transfer.

The amount of a chemical that is transferred to the fetus is dependent on lipid solubility, the degree of ionization, and the molecular weight. Lipophilic chemicals tend to diffuse across the placenta readily, while highly ionized compounds penetrate the placental membrane slowly. The molecular weight of a chemical affects placental transfer, with the larger molecules crossing the placental barrier less readily. Protein binding of a chemical or its metabolites will affect the rate and the amount transferred to the fetus. Exposure of fetal target tissues to chemical entities may also be influenced by metabolism in the placenta or the fetal liver (Juchau, 1972). Furthermore, maternal metabolism is a determining factor for the amount of drug to be transferred to the fetus. In extensive reviews, Juchau (1981, 1990) and Parke (1984) demonstrated that placental and/or fetal metabolism affects neonatal development and toxicity. The concentration of chemical in the fetal blood supply and elimination processes must also be considered in determination of fetal manifestations.

The physiological condition of the mother will affect placental transfer and fetal effects. During pregnancy there is a decrease in gastrointestinal absorption, which consequently delays the absorption of various drugs. In pregnancy there are marked increases in maternal body volumes and enhanced respiratory rates, which in turn affect the concentrations of chemicals reaching the fetus. In maternal diabetes the normal process of fetal lung maturation is delayed and manifested as infant respiratory distress syndrome (Bourbon & Farrell, 1985). The risk factor for fetal cerebral palsy is significantly increased in maternal thyroid disease or seizure patients (Newton, 1989). In the fetus itself genetic predisposition will affect neonatal development. The inability to metabolize phospholipid with subsequent accumulation results in Niemann-Pick and Tay-Sachs disease (Matsuzawa et al., 1977). Inborn errors of metabolism associated with enzymic deficiency are a source of fetal toxicity (Brent & Beckman, 1994). In the presence of an infectious disease with an associated rise in maternal body temperature, hepatic function in the mother can be compromised such that the use of theophylline in the treatment of asthma can produce severe toxicity in the fetus. During sustained fever the half-life of theophylline is significantly prolonged and can result in fetal cardiac and central nervous system (CNS) toxicity.

The list of factors in Table 1-1 should not imply that these are absolute with respect to influencing placental transfer and fetal effects. A combination of factors (multifactorial) can influence neonatal predisposition to chemical-induced toxicity. The lack of information on single-chemical exposure is self-evident, but even less is known about the interaction between two drugs or a drug and an environmental agent on fetal outcome. Certain neonates can be subjected to at least four different drug preparations in the course of therapy (Aranda et al., 1983). The manner in which drug–chemical interaction affects the fetus is a matter that requires intensive investigation. The adverse effects on the fetus of a mother who works in a smelting plant and smokes or an asthmatic mother who works in a plastics manufacturing firm remain to be established. In attempting to describe environmental or occupational pollutant interaction the consideration is primarily outdoors; however, indoor air pollution and fetal consequences must also be considered with respect to drug–chemical interactions. It should also be reiterated that the fetus and neonate are generally not "healthy" when there is an associated nutritional imbalance. The role of nutrition is a further complicating factor to be considered in multiple environmental, complex mixture exposures. In our current economic climate with excessive job losses and pressures, the interaction between stress and drugs/chemicals on fetal outcome will only be manifested in years to come, especially in infant behavioral development.

An integral component that is frequently overlooked is the effect of paternal chemical exposure on fetal outcome. Exposure of males to lead, morphine, ethanol, or caffeine has been shown to be associated with decreased birth weight and neonatal survival (Soyka & Joffe, 1980; Hill & Kleinberg, 1984). It was suggested that the observed fetal manifestations arising from paternal chemical exposure may result from direct damage to spermatozoa or an alteration in the intrauterine environment such that normal development cannot occur. It has been clearly established that paternal exposure to radiation resulted in malformed infants due to chromosomal aberrations. Exposure of females to formaldehyde, benzene, toluene, pesticides, etc. has been reported to produce menstrual disorders, increased rates of abortion, decreased fetal growth, and low-birth-weight infants (Schrag & Dixon, 1985). Although the precise mechanisms are not known, female reproductive functions including oogenesis, steroidogenesis, and ovulation can be affected by chemicals with subsequent fetal manifestations. It should be noted that damage to primary oocytes frequently results in cell death (Sonawane & Yaffe, 1988). However, genetically damaged oocytes can survive and become fertilized, with the embryo dying at the stage of implantation. Lead was found to decrease progesterone levels required for the blastocyte to enter the uterus for attachment to the endometrium, resulting in impaired implantation. Basler et al. (1976) indicated that a small fraction of oocytes, with chemically induced damage, reach the embryonic stage but subsequently develop malformations and congenital childhood disorders. Some chemicals such as diethylstilbestrol affect various stages including formation of abnormal sperm and delayed implantation through an antiestrogenic effect.

The stage of fetal development at the time of chemical exposure will have a bearing on fetal outcome. During the first week of development after fertilization, the embryo undergoes the process of cleavage and gastrulation. Exposure to drugs such as antimetabolites, ergot alkaloids, or diethylstilbestrol at this stage can result in termination of pregnancy (Roberts, 1986). Organogenesis is the next developmental stage, covering weeks 2–8 of gestation. Exposure to drugs including thalidomide, alcohol, lithium, phenytoin, and isotretinoin during this phase can result in serious structural abnormalities (Arena & Drew, 1986; Manson, 1986; Roberts, 1986). Chemicals such as cigarette smoke, heavy metals, or carbon monoxide may affect development during the remaining gestational period, ranging from 9 weeks to 9 months. Predominant effects are alteration in the differentiation of the reproductive and central nervous systems

(Roberts, 1986). Consequently, altered brain function and growth retardation are some of the principal adverse effects due to exposure at this stage.

There are four established classes of developmental toxicity in animals that can be directly compared with humans (Schardein & Keller, 1989). Intrauterine growth retardation, which is manifested as a low birth weight in humans, constitutes the first class of developmental toxicity, with an estimated frequency of 7%. Newton et al. (1987) demonstrated a higher perinatal death rate associated with intrauterine growth retardation. Similarly, in growth-retarded infants there is an increased incidence of congenital anomalies, neurological deficits, and learning disabilities (Christianson et al., 1981; Miller, 1981). It is of interest that the risk of intrauterine growth retardation is increased in a chronic hypertensive mother (Newton, 1989). It is well established that in chronic hypertension the mother would normally be receiving medication. However, the influence of this medication on the manifestation of intrauterine growth retardation or perinatal mortality is not known.

The second type of developmental toxicity is termed embryolethality and is manifested as spontaneous abortion with expulsion of product prior to the 20th week in humans, with an estimated frequency of 11–25% (Hook, 1981). It is believed that the predominant effect of toxicant exposure on development is spontaneous abortion and is closely linked with the presence of congenital malformations. The abortive process may thus be considered as a mechanism to terminate an abnormal conception (Haas & Schottenfeld, 1979). Death in the developing fetus beyond 20 weeks of gestation results in stillbirth or fetal death. When calculated per 1000 live births the frequency of stillbirth is 7–9. Finally, death occurring within the first 28 days of life is termed neonatal death with an estimated frequency of 6–8 per 1000 live births (Newton, 1989). Neonatal death can result from infection, an accident, or sudden infant death syndrome. However, Buehler et al. (1985) proposed that an increased percentage of neonatal death was merely a result of a postponement of lethal processes that were prevented from proceeding because of technological support systems. The role of chemicals in this latter phenomenon has yet to be established, but it should be noted that the responsiveness to toxicants in numerous cases is cumulative and delayed.

The third class of developmental toxicity is congenital malformation and in general is manifested as structural deformities. The congenital malformations are distinctly different from the behavioral or functional abnormalities. Congenital malformations are associated with intrauterine growth retardation but are also considered a major cause of perinatal mortality (Newton, 1989). Approximately 14–18% of infant deaths recorded were attributed to congenital malformations (Warkany, 1957; Newton et al., 1987).

The fourth class of developmental toxicity is termed functional disorder and excludes those abnormalities considered as outright terata (Schardein & Keller, 1989). In this category are included behavioral functions such as learning ability and neuromotor ability, as well as sensorimotor, reproductive, and respiratory functions. In the broadest sense impairment or deficits in any biological system would fall into this category. Functional disorders are one of the major causes of perinatal mortality (Newton, 1989) and are associated with congenital abnormalities (Smith & Bostian, 1964).

A perspective on the potential seriousness of the problems associated with drug use during pregnancy and the consequence to the fetus is gained from examination of the extensive tables outlined in the reviews by Lock and Kacew (1988) and Kacew (1993). The aim of this chapter is not to focus on a repetition of the numerous drugs listed and the fetal consequences but to raise awareness of seriousness of the potential adverse effects. Furthermore, some of the effects shown to occur in humans by one group of investigators could not necessarily be confirmed by other scientists. Despite these uncertainties, any reports of potential adverse effects of a drug on the unborn child should be taken into consideration by a physician. Thus, there should be a clear indication that the benefits derived by the mother from a specific drug greatly outweigh the risk to the fetus. Finally, these tables list self-administered drugs [over-the-counter (OTC) preparations] and nontherapeutic drugs (drugs of abuse). There is, unfortunately, less than adequate awareness of the dangers associated with the use of OTC drugs.

It has become increasingly apparent in recent years that fetal growth and development may also be subject to maternal exposure to industrial chemicals and environmental pollutants. The aim of this chapter is not to dwell on every chemical that is known to produce adverse fetal effects. More detailed information about specific compounds may be found in standard texts. The scope of industrial and/or environmental contaminants that are known to adversely affect the fetus is not vast in comparison to the number of chemicals currently present in the environment. However, in situations where environmental exposure was reported the degree of fetal injury extends from frank teratogenesis and spontaneous abortion to mental dysfunction, cardiac and renal anomalies, and more subtle biochemical and/or immunological effects. Although many of the published reports originate from animal studies where concentrations are excessively high, it should be noted that pregnant women are also exposed to these classes of chemicals. Hence, possible reactions to chemicals within the human population should not be dismissed lightly.

DRUGS, ENVIRONMENTAL AGENTS, AND CHILDREN

The pharmacokinetic principles applied in pediatric drug therapy are, in general, similar to those utilized for adults. However, data obtained in adult studies are not always applicable to rational therapy in infants or young children. The infant must be regarded as a distinct organism, and lack of appreciation of this fact could result in serious harm and potentially in death (Kacew, 1992). During infancy and childhood the body weight and composition are continuously changing such that pharmacodynamic aspects of drug therapy are not predictable. Certain drugs and chemicals are lipophilic and retained in body fat. As some children develop and "lose baby fat" there is less chemical retention, which is in contrast to the obese child.

A number of important characteristics exist that distinguish drug therapy in infants from adult medication protocols. For example, after intramuscular administration, drug absorption is partially dependent on blood flow in the muscle bed. Abnormal drug absorption following intramuscular injection can occur in premature infants, in whom muscle mass is small and blood flow to the musculature is poor. Examples of adverse effects attributed to altered drug absorption are the reactions of infants to cardiac glycosides and anticonvulsants (Cohen, 1984).

In the infant, absorption from the gastrointestinal tract of an orally administered drug differs from that in adults. In both adults and infants, the rate and extent of drug absorption depend on the degree of ionization, which, in turn, is influenced by pH. Within the first 24 h of life, gastric acidity increases rapidly, and this is followed by an elevation in alkalinity over the next 4–6 weeks (Roberts, 1986). These conditions result in drugs existing in the infant gastrointestinal tract in states of ionization other than might be observed in adults. Other factors that modify gastrointestinal drug absorption in the young infant as summarized by Roberts (1986) include an irregular neonatal peristalsis, a greater gastrointestinal tract surface to body weight ratio, and enhanced β-glucuronidase activity in the intestinal tract. The significance of the β-glucuronidase is that it releases drug-bound glucuronide to the free form and thus increases drug bioavailability. The bioavailability of drug is also related to the diurnal pattern. In general the urinary pH is acidic during sleep. In infants where the tendency is to sleep for a greater portion of the day, acid drug excretion would decrease in the presence of a low pH, and hence the bioavailability of sulfonamides would be enhanced during the sleep period. In contrast, the bioavailability of basic drugs would be reduced during the sleep phase. Differences exist in the organ distribution of drugs between newborns and adults. In the newborn, a higher percentage of body weight is represented by water, so extracellular water space is proportionally larger (Roberts, 1986). To initiate a receptor response, the distribution of drugs must occur predominantly in the extracellular space, so the concentration of drug reaching the receptor sites is higher in neonates. Furthermore, the ability of newborns to bind drugs in plasma is significantly less than in adults (Rane & Wilson, 1976). This is supported by the findings of Gorodischer et al. (1976), who reported a twofold increase in myocardial digoxin levels in infants as compared with adults. Similarly, a fivefold higher myocardium/serum ratio was subsequently found in infants administered digoxin in comparison with adults (Park et al., 1982). This again suggests that neonates could be expected to be more susceptible to the effects of drugs. Differences also exist with respect to drug-metabolizing enzymes. It has been clearly demonstrated that the drug inactivation rate is generally slower in newborns (Parke, 1984; Juchau, 1990). However, one cannot generalize about the drug-metabolizing capacity in newborns, as there is marked variability among infants and this capacity is highly dependent on the drug being examined (Rane et al., 1974). Furthermore, certain metabolic pathways exist uniquely in the neonate (Takkieddine et al., 1981; Parke, 1984). Although drug metabolism is a means of chemical inactivation, it is also a process utilized to form an active component. In an effort to obtain a therapeutically effective regimen this factor must be considered in light of the fact that hepatic drug-metabolizing enzyme capacity is age related. The appreciation of the role of drug metabolism in therapeutics suggests that further study is necessary to maximize the ability of the physician to calculate a desired drug dosage.

The knowledge of clearance rates is essential in consideration of initial as well as subsequent drug dosage in the patient. The ability of the neonate to eliminate drugs or chemicals via the kidney, the major excretion pathway, is significantly limited by the state of development of these organs. It is well established that the half-life of several antibiotics is prolonged in the neonatal period owing to a decreased glomerular filtration rate (Axline et al., 1967). However, the glomerular filtration rate reaches the level seen in adults by 5 months of age (West et al., 1948). In addition, the renal tubular secretory capacity increases during the first few months to attain adult values at 7 months of age. Consequently, an agent eliminated via the secretory renal pathway would have a threefold longer half-life in the infant. Aranda and Stern (1983) found that the clearance of phenytoin and phenobarbitone (phenobarbital) was rapid in the first week or two of age yet theophylline and caffeine clearance remained low until the age of 6 weeks. Leff et al. (1986) attributed the need for a greater dosage of phenytoin in infants compared with the adult to a result of increased metabolic clearance in the infant. Consideration of these factors indicates that the susceptibility and responsiveness of newborns to drug therapy are different from those of adults.

It has been estimated that over 150 different pharmacological agents are employed in neonatal therapy (Aranda et al., 1986). In addition, in a typical neonatal intensive care unit, it is estimated that every infant is exposed, on average, to five different medications. It should be stressed that in most cases the approval of drugs for adult therapy is not accompanied by adequate data to allow for use in children (Rane & Wilson, 1976). Despite the informational inadequacies the utilization of drugs in infants seems to be increasing. In their extensive studies, Aranda et al. (1982, 1983) reported that there were more than 30 different preparations that were administered at least once to the neonate. In addition, infants were subjected to polypharmacy as evidenced by at least four different drugs per baby in approximately 50% of the patient population. As indicated previously, the nutritional status of these infants is poor and susceptibility to adverse drug effects may be accentuated.

Drug utilization and consequent toxicity have also increased among the nonhospitalized pediatric population (Wiseman et al., 1987). Indeed, the prescribing habits of general practitioners were compared with the habits of hospital physicians for the condition of acute gastroenteritis in children. Over a 5-

year period Choonara et al. (1987) found that general practitioners prescribed significantly more medicines than did hospital physicians. When one considers that gastroenteritis is self-limiting, this indicates that drug utilization is still too high in nonhospitalized children. Failure to comply with the proper use of drugs can either lead to toxicity as a result of excess amount or to ineffective therapy from underdosing. Compliance failure in children can occur when parents are inadequately instructed, or where patients refuse to take the drugs as directed. Failure to comply is not restricted solely to the home but can occur in a hospital (Becker et al., 1972; Wilson, 1973). It is evident that the pharmacokinetic principles of drug utilization are dependent on proper pediatric compliance. The term "environment" need not necessarily apply to the presence of chemicals but also includes climate and seasons. It is of interest that Eskola and Poikolainen (1985) listed the number of drug poisonings in children aged 0–6 years and demonstrated a relationship to the season, with recurring peaks over a 3-year period. Various investigators have also reported on a correlation between season and drug poisoning (Basavaraj & Forster, 1982; Paulozzi, 1983). As antiasthmatic agents tend to be used to a greater extent in the spring compared to winter, one would expect a higher incidence of adverse reactions in the spring.

Unlike drugs, a chemical that has no direct therapeutic potential is not normally administered to infants. The potential for infants to be exposed to environmental agents is equivalent to adults in ambient air. Significant industrial pollution and photochemical smog in severe cases can produce respiratory irritation, edema, and hyperplasia. A concentration-dependent immunological response can occur as a consequence of exposure to certain chemicals. Because the lung cells of neonates are not fully developed, this population is more susceptible to the toxic actions of chemicals. As a general rule the potential risk of toxicity to chemicals or drugs is far higher in the neonate compared with the adult.

SUMMARY

The neonate and infant are unique organisms, and their responsiveness to drugs and other chemicals cannot be simply predicted from the known effects in adults. The consequences of maternal drug or chemical exposure can take years to become manifested in the offspring, so extreme caution should be used by women who believe themselves to be pregnant.

The remainder of this book is devoted to a discussion of the manner in which exposure to pharmacological and environmental agents during gestation and early life can affect maturation to adulthood.

REFERENCES

Aranda, J. V., and Stern, L. 1983. Clinical aspects of developmental pharmacology and toxicology. *Pharmacol. Ther.* 20:1–51.

Aranda, J. V., Collinge, J. M., and Clarkson, S. 1982. Epidemiologic aspects of drug utilization in a newborn intensive care unit. *Semin. Perinatol.* 6:148–154.

Aranda, J. V., Clarkson, S., and Collinge, J. M. 1983. Changing pattern of drug utilization in a neonatal intensive care unit. *Am. J. Perinatol.* 1:28–30.

Aranda, J. V., Chemtob, S., Laudignon, N., and Sasyniuk, B. I. 1986. Furosemide and vitamin E. Two problem drugs in neonatology. *Pediatr. Clin. North Am.* 33:583–602.

Arena, J. M., and Drew, R. H. 1986. Teratogenicity. In *Poisoning: Toxicology symptoms, treatments*, 5th ed., eds. J. M. Arena and R. H. Drew, pp. 997–1007. Springfield, IL: Charles C. Thomas.

Axline, S. G., Yaffe, S. J., and Simon, H. J. 1967. Clinical pharmacology of antimicrobials in premature infants. *Pediatrics* 39:97–107.

Basavaraj, D. S., and Forster, D. P. 1982. Accidental poisoning in young children. *J. Epidemiol. Community Health* 36:31–34.

Basler, A., Buselmaier, B., and Rohrborn, G. 1976. Elimination of spontaneous and chemically induced aberrations in mice during early embryogenesis. *Human Genet.* 33:121–130.

Basu, T. K. 1988. Nutritional factors and dispositions of pharmacological chemicals in prenatal and neonatal life. In *Toxicologic and pharmacologic principles in pediatrics*, eds. S. Kacew and S. Lock, pp. 17–40. Washington, DC: Hemisphere.

Becker, M. H., Drachman, R. J., and Kirscht, J. P. 1972. Predicting mothers' compliance with paediatric medical regimes. *J. Pediatr.* 81:843–854.

Bottoms, S. F., Kuhnert, B. R., Kuhnert, P. M., and Reese, A. L. 1982. Maternal passive smoking and foetal serum thiocyanate levels. *Am. J. Obstet. Gynecol.* 144:787–791.

Bourbon, J. R., and Farrell, P. M. 1985. Fetal lung development in the diabetic pregnancy. *Pediatr. Res.* 19:253–267.

Brent, R. L., and Beckman, D. A. 1994. The contribution of environmental teratogens to embryonic and fetal loss. *Clin. Obstet. Gynecol.* 37:646–670.

Buehler, J. W., Hogue, C. J. R., and Zaro, S. M. 1985. Postponing or preventing death? Trends in infant survival, Georgia, 1974 through 1981. *J. Am. Med. Assoc.* 253:3564–3567.

Choonara, I. A., Shoo, E. E., and Owens, G. G. 1987. Prescribing habits for children with acute gastroenteritis: A comparison over 5 years. *Br. J. Clin. Pharmacol.* 23:362–364.

Christianson, R. E., van den Berg, B. J., Milkovich, L., and Oechsli, F. W. 1981. Incidence of congenital anomalies among white and black live births with long-term follow-up. *Am. J. Public Health* 71:1333–1341.

Cohen, M. S. 1984. Special aspects of perinatal and pediatric pharmacology. In *Basic and clinical pharmacology*, ed. B. G. Katzung, pp. 749–755. Los Altos, CA: Lange.

Eskola, J., and Poikolainen, K. 1985. Seasonal variation and recurring peaks of reported poisonings during a 3-year period. *Human Toxicol.* 4:609–615.

Ginsburg, J. 1971. Placental drug transfer. *Ann. Rev. Pharmacol.* 11:387–408.

Gorodischer, R., Jusko, W. J., and Yaffe, S. J. 1976. Tissue and erythrocyte distribution of digoxin in children. *Clin. Pharmacol.* 19:256–263.

Haas, J. F., and Schottenfeld, D. 1979. Risks to the offspring from occupational exposure. *J. Occup. Med.* 21: 607–613.

Hill, L. M., and Kleinberg, F. 1984. Effects of drugs and chemicals on the foetus and newborn. *Mayo Clin. Proc.* 59:707–716.

Hook, E. B. 1981. Human teratologic and mutagenic markers in monitoring about point sources of pollution. *Environ. Res.* 25:178–203.

Juchau, M. R. 1972. Mechanisms of drug biotransformation reactions in the placenta. *Fed. Proc.* 31:48–51.

Juchau, M. R. 1981. Enzymatic bioactivation and inactivation of chemical teratogens. In *The biochemical basis of chemical teratogenesis*, ed. M. R. Juchau, pp. 63–94. Amsterdam: Elsevier North Holland.

Juchau, M. R. 1990. Fetal and neonatal drug biotransformation. In *Drug toxicity and metabolism in pediatrics*, ed. S. Kacew, pp. 15–34. Boca Raton, FL: CRC Press.

Kacew, S. 1992. General principles in pharmacology and toxicology applicable to children. In

Similarities and differences between children and adults: Implications for risk assessment, eds. P. S. Guselian, C. J. Henry, and S. S. Olin, pp. 24–34. Washington, DC: ILSI Press.

Kacew, S. 1993. Adverse effects of drugs and chemicals in breast milk on the nursing infant. *J. Clin. Pharmacol.* 33:213–221.

Kacew, S. 1994a. Current issues in lactation: Advantages, environment, silicone. *Biomed. Environ. Sci.* 7:307–319.

Kacew, S. 1994b. Fetal consequences and risks attributed to the use of over-the-counter (OTC) preparations during pregnancy. *Int. J. Clin. Pharmacol. Ther.* 32:335–343.

Leff, R. D., Fischer, L. J., and Roberts, R. J. 1986. Phenytoin metabolism in infants following intravenous and oral administration. *Dev. Pharmacol. Ther.* 9:217–223.

Lock, S., and Kacew, S. 1988. General principles in pediatric pharmacology and toxicology. In *Toxicologic and pharmacologic principles in pediatrics,* eds. S. Kacew and S. Lock, pp. 1–15. Washington, DC: Hemisphere.

Mahaffey, K. R. 1980. Nutrient-lead interactions. In *Lead toxicity,* eds. R. L. Singhal and J. A. Thomas, pp. 425–460. Baltimore, MD: Urban and Schwarzenberg.

Manson, J. M. 1986. Teratogens. In *Casarett and Doull's toxicology: The basic science of poisons,* 3rd ed., eds. C. D. Klaassen, M. O. Amdur, and J. Doull, pp. 195–220. New York: Macmillan.

Matsuzawa, Y., Yamamoto, A., Adachi, S., and Nishikawa, M. 1977. Studies on drug-induced lipidosis. *J. Biochem.* 82:1369–1377.

Miller, H. C. 1981. Intrauterine growth retardation. An unmet challenge. *Am. J. Dis. Child.* 135:944–948.

Newton, E. R. 1989. The foetus as a patient. *Med. Clin. North Am.* 73:517–540.

Newton, E. R., Kennedy, J. L., and Louis, F. 1987. Obstetric diagnosis and perinatal mortality. *Am. J. Perinatol.* 4:300–304.

Park, M. K., Ludden, T., Arom, K. V., Rogers, J., and Oswalt, J. D. 1982. Myocardial versus serum digoxin concentrations in infants and adults. *Am. J. Dis. Child.* 136:418–420.

Parke, D. V. 1984. Development of detoxication mechanisms in the neonate. In *Toxicology and the newborn,* eds. S. Kacew and M. J. Reasor, pp. 3–31. Amsterdam: Elsevier.

Paulozzi, L. J. 1983. Seasonality of reported poison exposures. *Pediatrics* 71:891–893.

Rane, A., and Wilson, J. T. 1976. Clinical pharmacokinetics in infants and children. *Clin. Pharmacokinet.* 1:2–24.

Rane, A., Garle, M., Borga, O., and Sjoqvist, F. 1974. Plasma disappearance of transplacentally transferred phenytoin in the newborn studied in mass fragmentography. *Clin. Pharmacol. Ther.* 15:39–45.

Roberts, J. R. 1986. Developmental aspects of clinical pharmacology. In *The scientific basis of clinical pharmacology,* ed. R. Spector, pp. 153–170. Boston: Little, Brown.

Schardein, J. L., and Keller, K. A. 1989. Potential human developmental toxicants and the role of animal testing in their identification and characterization. *CRC Crit. Rev. Toxicol.* 19: 251–339.

Schrag, S. D., and Dixon, R. L. 1985. Reproductive effects of chemical agents. In *Reproductive toxicology,* ed. R. L. Dixon, pp. 301–319. New York: Raven Press.

Smith, D. W., and Bostian, K. E. 1964. Congenital anomalies associated with idiopathic mental retardation. *J. Pediatr.* 65:189–196.

Sonawane, B. R., and Yaffe, S .J. 1988. Drug exposure in utero: Reproductive function in offspring. In *Toxicologic and pharmacologic principles in pediatrics,* eds. S. Kacew and S. Lock, pp. 41–65. Washington, DC: Hemisphere.

Soyka, L. F., and Joffe, J. M. 1980. Male-mediated drug effects on offspring. *Prog. Clin. Biol. Res.* 36:49–66.

Takkieddine, F. N., Tserng, K. Y., King, K. C., and Kalhan, S. C. 1981. Postnatal development of theophylline metabolism in preterm infants. *Semin. Perinatol.* 5:351–358.

Warkany, J. 1957. Congenital malformations and paediatrics. *Pediatrics* 19:725–733.

West, J. R., Smith, H. W., and Chasis, H. 1948. Glomeraular filtration rate, effective renal blood flow, and maximal tubular excretory capacity in infancy. *J. Pediat.* 32:10–18.

Wilson, J. T. 1973. Compliance with instructions in the evaluation of therapeutic efficacy: A common but frequently unrecognized major variable. *Clin. Pediatr.* 12:333–340.

Wiseman, H. M., Guest, K., Murray, V. S. G., and Volans, G. N. 1987. Accidental poisoning in childhood: A multicentre study. 1. General epidemiology. *Hum. Toxicol.* 6:293–301.

Chapter 2

Retinoids in Abnormal and Normal Embryonic Development

Joan Creech Kraft and Calvin C. Willhite

The origin of the term vitamin A can be traced to the independent investigations of McCollum and Davis (1913) and Osborne and Mendel (1913). McCollum actually proposed the terms *fat-soluble A* for the factor or substances in butter that alleviate night blindness and abnormal keratinization and *water-soluble B* for the factor or substances in foods that ameliorate the clinical signs of beri beri; the A and B were simply assigned because the former was the first extracted from foods with solvent and the latter extracted second with water. However, as early as 400 BC, the ancient Chinese used vitamin A-rich foods in treatment of night blindness; Hippocrates recommended raw ox liver dipped in honey, and by 1816, clinical recognition of diet in the etiology of xerophthalmia had been made (Magendie, 1816). By the 1920s, the term "fat-soluble A vitamin" (stemming from the earlier concept of *vitamine* as a class of nutrient distinct from carbohydrate, fat, and protein) had come into common use (Wolbach & Howe, 1925). While today vitamin A deficiency is considered relatively rare in developed countries, endemic vitamin A deficiency is commonplace in poor countries. In Tibet, for example, the children of entire villages are afflicted with xerophthalmia; as gestation progresses, pregnant women often experience night blindness. There is an elevated incidence of neonatal mortality, which can be reduced dramatically by bolus vitamin A administration. For decades, international health organizations have carried out mass vitamin A depot injections in Southeast Asia, India, and other desperately poor countries (Underwood, 1993).

This work was supported in part by NIEHS grant ES05861. We thank Dr. Eva Stimac for her helpful criticism of our manuscript.

In 1960, the International Union of Pure and Applied Chemistry (IUPAC) adopted the terms retinol (vitamin A_1), retinal, and retinoic acid (RA) to standardize the literature. Nonetheless, the generalized term vitamin A persists in contemporary literature despite the marked differences in physiologic function and pharmacological action of the naturally occurring polyene congeners. Today the retinoids, a broad family of diverse chemical structures, represents not only those compounds that possess physiologic actions associated with the historical vitamin A (Wolf, 1984), but also includes compounds that bind to and transactivate the cognate nuclear receptors (discussed later) (or in the course of biotransformation give rise to metabolites that do so) and includes their physiologic (e.g., the retinyl esters) or synthetic pharmacological precursors (e.g., etretinate), exclusive of the carotenoids. Whether those congeners that have no detectable pharmacologic actions in vivo or in vitro or that bind—but do not transactivate—the nuclear receptors can be considered retinoids can be debated. Retinoids participate in the control of exceedingly diverse biologic activities including the cis- and trans-isomerization of retinals of visual purple in the retina, epithelial keratinization, osteoclast and osteoblast activities, fatty acid metabolism, immunologic activity, and embryogenesis (reviewed in Sporn et al., 1994). The primary focus of this chapter is the role of retinoids in normal and abnormal embryogenesis.

NORMAL EMBRYOGENESIS

In order to understand the effects of RA on embryonic development, the following short description of normal events during early embryogenesis should be kept in mind. One fundamental puzzle is the mechanism whereby the early sphere of cells (morula) comes to be shaped into increasingly complex levels of organization: first, through simple cell division; second, through cell differentiation; third, to functional division of these many different cell types into all of the diverse organs of the body, all placed in their proper location. In short, at the beginning how does an embryo know top from bottom, left from right, and front from back? Retinoids are definite keys that control differentiation and pattern organization in embryos or "formation of the body plan" where cells are informed where they are ("positional values") and given information about where they are supposed to go (Wolpert, 1969). The establishment of the anteroposterior axis is quite similar in all vertebrates (see Bellairs et al., 1986; Slack & Tannahill, 1992; Lawson & Pederson, 1992), and it is specified early in development when the embryo undergoes several complex cellular rearrangements known as gastrulation and neurulation. The primitive streak is the first manifestation of a structural feature that delineates the anteroposterior polarity in the embryo. It begins to form at the posterior end of the embryo and elongates anteriorly along the midline. During formation of the streak, cells from the ectodermal layer of the embryo ingress through the streak and then stream laterally, giving rise to the mesoderm and endoderm. After the streak has reached its most anterior extension, which is termed the node (organizer), it regresses toward the posterior pole of the embryo. As the node regresses, adjacent mesodermal and ectodermal tissues form axial structures. Along the midline, mesoderm gives rise to the notochord, and the more lateral mesoderm is organized into the segmented somites. A band of ectoderm along the midline forms the neural plate, the primordium of the nervous system. The neural plate thickens and elevates to a U-shaped structure that fuses in the midline to form the neural tube, the primordium of the fore-, mid-, and hindbrain and spinal cord. The closing neural tube releases clusters of cells known as the neural crest. As the neural crest cells migrate into the future head and neck region, the branchial apparatus (pharyngeal arches) forms. Most congenital malformations induced by retinoids of the head and neck originate during neural crest migration, during formation of the branchial apparatus and its development into adult derivatives. However, alteration of any number of these early events can cause embryonic malformation or death.

VITAMIN A AND TERATOGENESIS

Isotretinoin (Accutane; 13-*cis*-retinoic acid; 13-*cis*-RA) was first confirmed as a human teratogen in 1985, but as early as 1983, the initial reports of isotretinoin embryopathy had appeared (Rosa, 1983). The characteristic features of isotretinoin embryopathy involve malformations of the heart, thymus, face, jaw, ears, palate (Figures 2-1 to 2-3), and brain (Lammer et al., 1985). The risk of major malformations approximates 25% among those fetuses that survive to 20 weeks of gestation. The maternal oral administered dose associated with this elevated risk ranges between 0.5 and 1.5 mg/kg/day. Even only brief exposure (3 days) during the most sensitive stage of human gestation (1 month) has been sufficient to induce retinoid embryopathy (Hersh et al., 1985). By 1990, the U.S. Food and Administration (FDA) had received detailed reports on 89 affected infants (Rosa, 1990), and by 1993, 94 cases had been confirmed (Schardein, 1993). These data do not include spontaneous or elective abortions associated with isotretinoin exposures, and do not include cases that went unreported to the U.S. FDA. To prevent isotretinoin use during pregnancy, there is a requirement for a signed consent form, acknowledgment of receipt of verbal and written warnings on teratogenic risks, a negative serum pregnancy test, and the onset of therapy must be on menses day 2–3. By 1984 the related aromatic retinoid etretinate was recognized as a human teratogen (Happle et al., 1984).

The obvious physical features of isotretinoin embryopathy have been well documented (Braun et al., 1984; Willhite et al., 1986). Many have considerable motor deficits and are unable to walk or sit (Adams & Lammer, 1991). As was the case with thalidomide (Jorgensen et al., 1964), among the most prevalent malformations in retinoid embryopathy are those of the ears. Microtia or anotia, which is usually asymmetric with low-set tag remnants of pinnae and absent/stenotic external auditory canals, is common (Figure 2-3). Autopsy of neonates

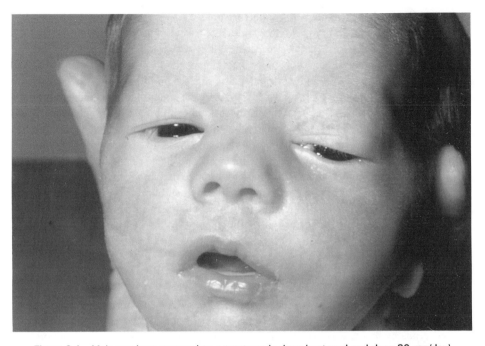

Figure 2-1 Male newborn exposed to a teratogenic dose (maternal oral dose 80 mg/day) of 13-*cis*-RA from day 0 to day 42 of gestation. This child exhibits hypertelorism, a depressed nasal bridge, left facial paralysis, and is afflicted with a cleft of the primary and secondary palates. This child was born at 38 weeks gestation (2.8 kg) and experienced chronic respiratory distress, and he required both tracheostomy and gastrostomy. The chronic cyanosis and respiratory insufficiency were due in large measure to the ventricular septal defect, aortic and pulmonary stenosis, and the dysplastic pulmonary arch. (Reproduced with permission from C. C. Willhite, R. M. Hill, and D. W. Irving, *J. Craniofac. Genet. Develop. Biol.* Suppl. 2:193–209, 1986, © 1996 Munksgaard International Publishers Ltd., Copenhagen, Denmark.)

and children who died between 19 weeks and 3 years of age have confirmed middle and inner ear malformations, anomalies in the sphenoid and temporal bones, and defective innervation, all of which contribute to the congenital deafness (Burk & Willhite, 1992). These malformations are all consequences of retinoid-induced pathology of the pharyngeal arches and fundamental disturbance in neural crest migration and function. The two basic types of inner ear defects induced by retinoids [as reproduced in mouse (Jarvis et al., 1990) and hamster (Burk & Willhite, 1992)] are Michel aplasia and the Mondini–Alexander defect. In the case of Michel aplasia, there is a complete lack of, or vestigial, inner ear, and petrous temporal development. The Mondini–Alexander defect is comprised of flattened cochlea, limited to basal coil and hypoplastic vestibular structures, and the inner sensory epithelium is limited to the sacculovestibular cavity (Figures 2-3 to 2-6). Children with isotretinoin-induced Mondini defect

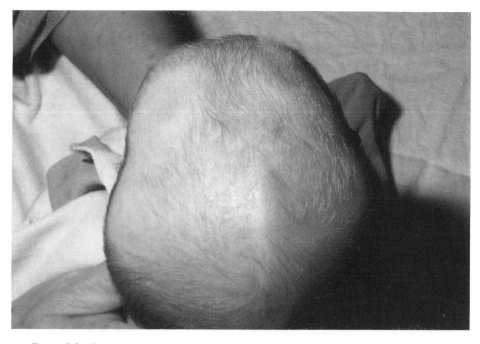

Figure 2-2 Superior view of the child shown in Figure 2-1. Note the triangular skull, narrow forehead, and prominent occiput. (Reproduced with permission from C. C. Willhite, R. M. Hill, and D. W. Irving, *J. Craniofac. Genet. Develop. Biol.* Suppl. 2:193–209, 1986, © 1996 Munksgaard International Publishers Ltd., Copenhagen, Denmark.)

(the severity of which seems to be correlated with the severity of the external aural malformation) have a cochlea with fewer turns than normal, near complete absence of cochlear neurons, and an abnormally large utricle and saccule. The most severe manifestation of inner ear terata (Figure 2-5) is a complete arrest of development at the otocyst stage with or without sensory innervation. Three-dimensional reconstruction of the inner ears shows a range of malformation (Figure 2-7) even among littermates. The malformations of the auditory system are generally bilateral in animals, but depending upon the particular patient, this may or may not be the case in humans.

The pathogenesis of retinoid-induced external, middle, and inner ear terata has, as its basis, two distinct avenues; the first, production of anomalies in the embryonic structures that give rise to the viscerocranium, and the second, an apparent direct toxicity in the neuroepithelium of the rhombencephalon. In humans, the facial skeleton begins to form on day 22 when the first of the 5 paired pharyngeal arches appears. The pharyngeal arches, often referred to as branchial (*branchial* is from the Greek meaning gill) arches, are complicated ancient structures corresponding to the gill bars 1, 2, 3, 4, and 6 in jawed fishes. This apparatus, consisting of external ectoderm, internal endoderm (gut lining),

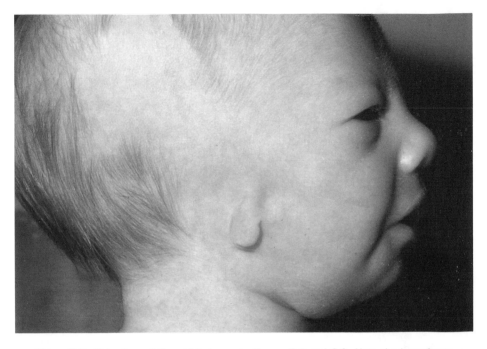

Figure 2-3 Side view of the child shown in Figures 2-1 and 2-2. Note the imperforate auditory canal, the dysplastic pinnae remnants set at the angle of the mandible, and micrognathia. This child also had a prolonged latency in auditory brainstem response. (Reproduced with permission from C. C. Willhite, R. M. Hill, and D. W. Irving, *J. Craniofac. Genet. Develop. Biol.* Suppl. 2:193–209, 1986, © 1996 Munksgaard International Publishers Ltd., Copenhagen, Denmark.)

covering lateral plate mesoderm, and ectomesenchymal cells, is of neural crest origin. The first arch (with maxillary and mandibular processes having those elements of the upper and lower jaw) is formed from neural crest migrating from the level of the mesencephalon and metencephalon, and it gives rise from the maxillary process to—among other structures—the maxilla, and zygomatic and squamous temporals, through membranous ossification and endochondral ossification to the incus. The lower (mandibular) process of the first arch gives rise to the jaw through membranous ossification of Meckel's cartilage and by endochondral ossification to the malleus. The second pharyngeal arch condenses to form Reichert's cartilage, which undergoes endochondral ossification to form the stapes. In humans, the second and third arch appear on day 24, comprised of neural crest cells migrating from the level of the embryonic myelencephalon. Both the first and second arch contribute to the external ear, and the cleft between the two gives rise to the external auditory canal. The branchial apparatus of isotretinoin-treated fetuses is obviously hypoplastic (Irving et al., 1986), resulting in the anomalies of the malleus, incus, stapes, and related structures (temporals, Meckel's and Reichert's cartilage) (Jarvis et al., 1990). The severe

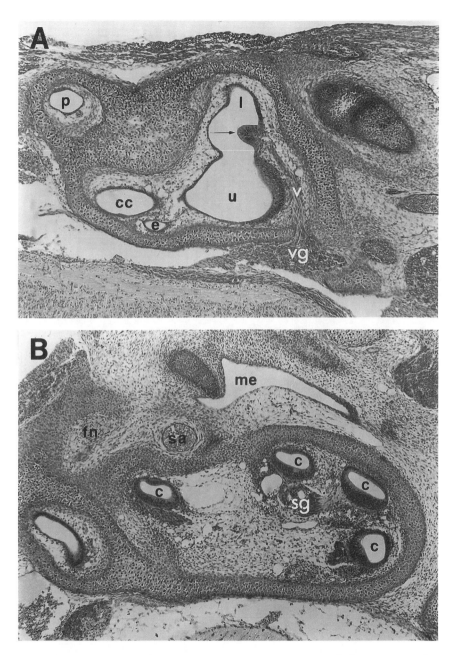

Figure 2-4 (A) Low magnification of a section through the vestibular apparatus of a control hamster fetus. The lateral semicircular duct (1) is entering the utricle (u). Nerve fibers of the vestibular nerve (v) are visible between the vestibular ganglion (vg) and the crista of the lateral semicircular duct (arrow); posterior semicircular duct (p), crus communis (cc), endolymphatic duct (e). x80. (B) Low magnification of a section through the cochlea of a control fetus; cochlear duct (c), spiral (cochlear) ganglion and nerve (sq), middle ear cavity (me), stapedial artery (sa), facial nerve (fn). ×80 (Reproduced with permission from D. T. Burk and C. C. Willhite, *Teratology* 46:147–157, © 1992 Wiley-Liss, Inc., a subsidiary of John Wiley & Sons, Inc.)

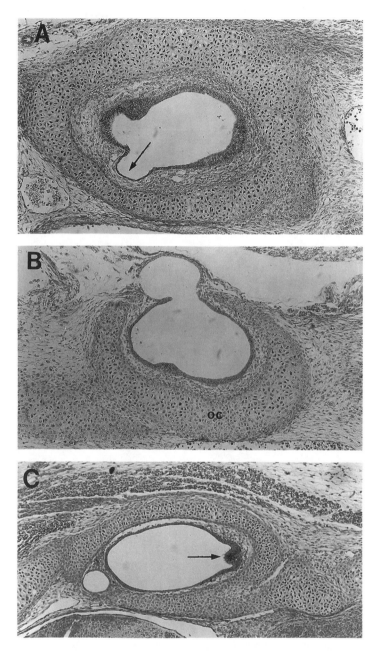

Figure 2-5 Sections through the severely malformed inner ears of isotretinoin-treated hamster fetuses. Anterior is to the right and lateral toward the top. x100 (A) Simple sac-like otocyst with a small epithelial projection on the medial side (arrow). (B) Inner ear is represented by an epithelial sac with a lateral projection herniating through the otic capsule (oc). (C) Otocyst contains tissue with features of a sensory area (arrow) but lacking innervation. (Reproduced with permission from D. T. Burk and C. C. Willhite, *Teratology* 46:147–157, © 1992 Wiley-Liss, Inc., a subsidiary of John Wiley & Sons, Inc.)

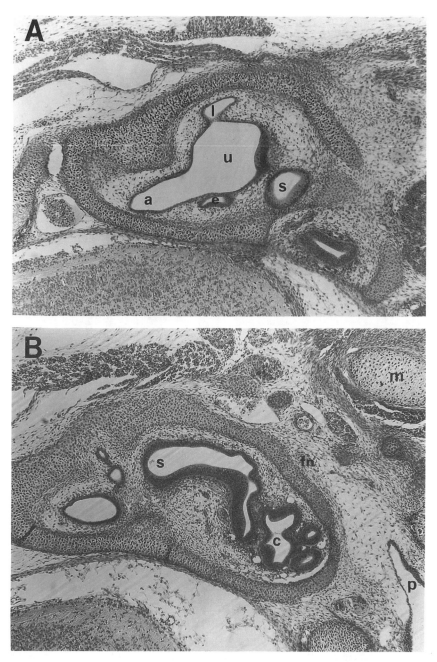

Figure 2-6 (A) Low magnification of a section through the inner ear (vestibular apparatus and a portion of the cochlea) of an isotretinoin-treated hamster fetus. Utricle (u), lateral extension from utricle (1); anterior semicircular duct (a); endolymphatic duct (e); saccule (s). ×80. (B) Low magnification of a section through the cochlear duct (c) is small and tight but at least two coils are evident; saccule (s), pharynx (p), facial nerve (fn), Meckel's cartilage (m). ×80. (Reproduced with permission from D. T. Burk and C. C. Willhite, *Teratology* 46:147–157, © 1992 Wiley-Liss, Inc., a subsidiary of John Wiley & Sons, Inc.)

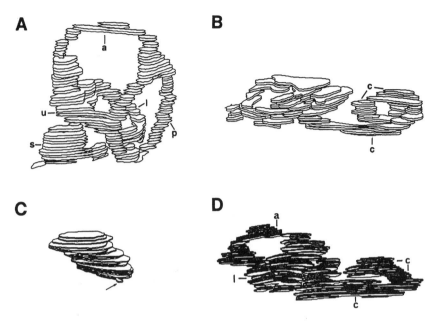

Figure 2-7 (A) Reconstruction of the vestibular apparatus of a control hamster fetus observed from the medial side and 15 degrees above the horizontal plane. Anterior is to the left; anterior semicircular duct (a), posterior semicircular duct (p), lateral semicircular duct (1), utricle (u), saccule (s). (B) Reconstruction of the cochlear duct (c) of a control hamster fetus viewed from anterolateral and 15 degrees above the horizontal plane. At least one and one-half turns in the cochlear duct are apparent. (C) Reconstruction of a severely malformed inner ear from an isotretinoin-treated hamster fetus viewed from the medial side and 15 degrees above the horizontal plane. Anterior is to the right. The entire inner ear is comprised of a sac-like epithelial structure (otocyst). A small projection of epithelium is present (arrow). (D) Reconstruction of a moderately malformed inner ear of an isotretinoin-treated hamster fetus viewed from anterolateral and 15 degrees below the horizontal plane. The anterior semicircular duct (a) is complete and there is a broad lateral projection (1) from the utricle in a location suggesting the lateral semicircular duct. The cochlear duct (c) in this particular fetus was reduced in length and makes approximately one complete turn. (Reproduced with permission from D. T. Burk and C. C. Willhite, *Teratology* 46:147–157, © 1992 Wiley-Liss, Inc., a subsidiary of John Wiley & Sons, Inc.)

malformations of the inner ear and the complete absence of 8th cranial nerve innervation could reflect a completely different site of action, either direct cyto-toxicity within the otocyst itself (Figure 2-5), or the equally plausible disruption of rhombencephalic segmentation or toxicity within the neuroepithelium of that structure and subsequent interference with the necessary inductive influence of the hindbrain on the otic vesicle (Burk & Willhite, 1992).

The concordance between features of retinoid embryopathy in humans and laboratory animals is extraordinary. Just as the obvious physical defects (Figures 2-1 to 2-3) are so similar, so are the functional deficits (behavioral terata) induced by prenatal retinoid exposures (Adams, 1993). In the follow-up evaluation of 31 children at age 5 years (±3 months) who had been exposed to isotretinoin during the first 60 days of gestation, more than one-third of the cohort were classified as either mentally retarded or functional only at full-scale IQ of 70–80 (Adams & Lammer, 1991). While there was an increased prevalence of frank mental retardation among children with major congenital malformation, these studies confirm the hypothesis that prenatal retinoid exposure can induce neuropsychologic deficits in the absence of anatomically detectable pathology of the central nervous system. Having one or more major congenital malformations of the brain (cerebellar hypoplasia, absent/malformed vermis, inferior olive malformed, dilated ventricles, focal cortical agenesis, hydrocephalus, absent aqueduct, optic nerve hypoplasia) was invariably associated with borderline or frank mental retardation (Lammer et al., 1985; Adams & Lammer, 1991).

The central nervous system (CNS) malformations and the terata of the inner and middle ear (Figures 2-4 to 2-7) can be traced to a fundamental interference with the tissue-specific expression of the homeobox genes (Hunt et al., 1991).

MOLECULAR MECHANISM OF RETINOIDS/CELLULAR RETINOL AND RETINOIC ACID BINDING PROTEINS

There are two cellular retinol-binding proteins, CRBP and CRBP II (for review see Chytil & Ong, 1978; Willhite, 1993). CRBP is widely distributed in almost all organs, whereas CRBP II is more restricted and found only in small intestine enterocytes (Levin et al., 1987). CRBP II participates in retinal/retinol conversion and subsequent esterification. In the liver, as well as target cells, CRBP plays an important role in retinoid metabolism, allowing retinol to interact with enzymes by a direct protein–protein mechanism. CRBP directs the esterification/deesterification of retinol and subsequent oxidation of retinol to retinal and RA (Napoli et al., 1991, 1993; Napoli, 1993; Ong et al., 1994). Metabolic activation of retinol to all-*trans*-RA involves two dehydrogenase enzymes, one for retinol, alcohol dehydrogenase (ADH), and one for retinal, aldehyde dehydrogenase (ALDH) (Leo et al., 1989; M.-O. Lee et al., 1991). ADH may be directly (feedback) regulated as induced by RA (Deuster et al., 1991). All-*trans*-RA induces the expression of CRBP (Smith et al., 1991). A retinoic acid response element (RARE) from the promoter region of the CRBP gene has been identified (Husmann et al., 1992), and it is activated by a nuclear retinoic acid receptor (RAR/RXR) heterodimer (discussed later). This implies that 9-*cis*-RA and/or all-*trans*-RA may be involved in the transcription of a protein that is involved in controlling the bioconversion of retinol to RAs. In mouse embryos, CRBP is expressed in tissues most vulnerable to vitamin A deficiency, that is, the heart and eye (Ruberte et al., 1991), suggesting that CRBP-mediated RA synthesis occurs in those tissues that require elevated levels to develop normally (Morriss-Kay, 1993).

In 1978, cellular retinoic acid binding protein (CRABP) was discovered (Chytil & Ong, 1978). It is a cytoplasmic protein that binds all-*trans*-RA with high affinity. Originally, many investigators believed that this was the receptor for all-*trans*-RA. There have been, however, no studies to show that CRABP can regulate transcription or play specific roles in nuclear RA signaling. CRABP has often been called a transport protein, but cells lacking this protein do not require it for transport (Haussler et al., 1984; Mangelsdorf et al., 1990). There are at least two types of CRABPs, CRABP I and II, and these are believed to participate in RA metabolism and to control levels of free RA in the cytoplasm (reviewed in Morriss-Kay, 1993). All-*trans*-RA can induce the expression of CRABP (Astrom et al., 1991; MacGregor et al., 1992; Durand et al., 1992). There is a clearly defined spatiotemporal pattern for these binding proteins during morphogenesis (Ruberte et al., 1991, 1992). CRABP I is believed to play an important role in retinoid metabolism, allowing its ligands to interact with cytochrome P-450 enzymes (Napoli et al., 1991, 1993; Napoli, 1993). Studies of Fiorella and Napoli (1991) and Boylan and Gudas (1992) showed that CRABP I enhanced the metabolism of RA. More recently, it has been suggested that the lower concentration of RA anteriorly in the chick limb bud may be due to enhanced RA metabolism facilitated by higher CRABP I concentrations in anterior tissue (Scott et al., 1994). In spite of an enormous volume of information regarding RA and its cellular binding proteins (reviewed in Dencker et al., 1991; Napoli et al., 1993; Napoli, 1993; Petkovich et al., 1992; Morriss-Kay, 1993; Willhite, 1993), there is still a lack of knowledge of the precise role CRABPs may be playing in teratogenicity. CRABP type I is present in chick wing (along with its cousin CRABP type II), and the presence of this protein enhances the metabolism of all-*trans*-RA there (Maden et al., 1988; for review, see Maden, 1994b). The level of CRABP I gene expression normally controls the local amount and nature of the RA metabolites produced; thus, CRABP I function is intimately related to control of local concentrations of free acidic retinoid. The function of CRABP II (which is present in much higher concentrations than CRABP I) is not known, but it is a candidate for an intracellular transport mechanism (shepard function) (Morriss-Kay, 1993). Willhite (1993) concluded that most teratogenic retinoids bound embryonic CRABP, but that in vitro binding affinities of teratogenic retinoids that competed for embryonic CRABP failed to correlate with relative teratogenic potency. More recent data provide evidence that CRABPs have no direct role in the biochemical mechanism of retinoid teratogenesis in that acidic teratogenic retinoids exist that do not bind this protein (Willhite et al., 1996). Mice deficient in CRABP II or in both CRABP I and CRABP II were essentially normal (Lampron et al., 1995). Single- and doublemutant CRABP mice were not more sensitive to RA excess treatment in utero; while the CRABPs may be dispensable during mouse development, their function appears to be maintenance of normal intracellular RA homeostasis and to maintain physiological levels of RA under conditions of limited supply (Lampron et al., 1995). Cells that express CRABP may be uniquely sensitive to precise concentrations of acidic retinoids, using CRABP to control free retinoid concentrations and to transfer RA to microsomal enzymes for catabolism (Napoli et al., 1993).

MOLECULAR MECHANISM OF RETINOIDS/RETINOID
NUCLEAR RECEPTORS

In 1987, Giguere et al. and Petkovich et al. independently isolated a human receptor cDNA that encoded the first RA-activated transcription factor (for reviews see Petkovich, 1992; Giguere, 1994). This receptor protein, associated with the nucleus, activates the transcription of target genes with all-*trans*-RA in vitro at concentrations effective in vivo (Evans, 1988). RA-bound RAR dimers (Beato, 1989) act by binding RA response elements (RAREs) in the promotor of target genes. This discovery provided the first mechanistic pathway in the biochemical pharmacology of all-*trans*-RA. Three different subtypes (RARα, RARβ, RARγ) exist in mammals, birds and amphibians (for review see Mangelsdorf, 1994). In embryos, RARα, β, and γ show very specific spatial and temporal patterns of expression, providing the potential for great variation in the response to all-*trans*-RA in different locations and at different developmental stages (Ruberte et al., 1991).

Retinoid nuclear receptors are part of the steroid/thyroid hormone receptor superfamily. These proteins control specific gene expression through a ligand-specific binding paradigm. Following receptor–ligand interaction, these proteins (acting as dimers under physiologic conditions) bind to short DNA segments (termed hormone response elements, HRE) that are located upstream of the target gene. Binding of the receptor complex to the cognate HRE activates transcription. Interaction of these trans-acting proteins on cis-acting DNA elements can result in repression or activation of biological pathways. The genes encoding the receptors have five regions termed A/B, C, D, E, and F. Region E is the ligand-binding domain; region C is the 66–68 amino acid zinc finger DNA binding domain. There is a high degree of homology across species; for example, 21 of the region C amino acids never vary and there is a nearly 90% concordance between humans and birds. The remaining regions either are structural components (the region D hinge) or vary between receptors of the same family (region A/B) or may not appear on all of the receptors (region F) (Giguere et al., 1987).

Although it would be tempting to speculate that retinoid-induced teratogenicity is mediated in toto by the RARs, the story is far more complicated, since it was shown that in transgenic mouse lines in which single RAR isoforms had been disrupted by gene targeting, there were no apparent effects on development or viability (Lohnes et al., 1992; Lufkin et al., 1993). However, another class of retinoid nuclear receptors, the retinoid X receptors (RXR), has been identified (Mangelsdorf et al., 1990), and it is now known that the normal high-affinity ligand for this receptor is 9-*cis*-RA, an isomer of all-*trans*-RA (Heyman et al., 1992; Levin et al., 1992). 9-*cis*-RA binds and activates not only RXRs, but also RARs. It has also been reported that 9-*cis*-didehydro-RA (9-*cis*-dd-RA) can likewise bind and activate the RARs and RXRs (Allenby et al., 1993). The RXRs can function independently via homodimer formation and activate a specific subset of RAREs (Zhang et al., 1992a), but the RARs do not appear to function independently. In order for RARs to bind RAREs and subsequently activate transcription of genes, heterodimerization with RXR is required (Lehmann et al., 1992). The presence of two receptor systems (RXR and RAR) implies that there are target genes specifically responsive to each combination of receptors. In order to regulate the target genes, both the receptor and the ligand must be present within that same cell.

RXRs also heterodimerize with thyroid receptors (TRs), the vitamin D3 receptor (VDR), and the peroxisome proliferator-activated receptor (PPAR) (Yu et al.,1991; Leid et al., 1992; Kliewer et al., 1992a, 1992b; Zhang et al., 1992b, 1992c; Poellinger et al., 1992). The principal metabolite of the plasticizer di(2-ethyhexyl) phthalate (DEHP) is mono(2-ethylhexyl) phthalate; the latter is held responsible for the weak DEHP carcinogenic activity as a peroxisome proliferator, and the metabolite is a ligand for peroxisome proliferator-activated receptor alpha (PPARα) (Gottlicher et al., 1992; Dreyer et al., 1992), a receptor analogous to the human hNUC1 (Schmidt et al., 1992). These receptors have their highest expression in the liver (Issemann & Green, 1990) and are involved in normal utilization of fatty acids; the PPARα/RXRα dimer (Kliewer et al., 1992b) recognizes a specific HRE that is located upstream of the peroxisomal acyl-CoA (CoA is coenzyme A) oxidase gene. Peroxisome proliferator response elements (PPRE) are also located in the regulatory regions of other genes, including CYPA6 (Muerhoff et al., 1992) and peroxisomal bifunctional enzyme (Zhang et al., 1992c), all of which have tandem repeats of a common heterodimer TGACCT binding half-site, separated with single base pair spacing. High concentrations of dietary fatty acids activate the PPARα (Gottlicher et al., 1992), leading to hepatic peroxisome proliferation (Osmundsen et al., 1991). Chronic high exposures to compounds like DEHP appear to induce inappropriate gene expression—including genes normally regulated in tandem with the promiscuous RXR (Kliewer et al., 1992b)—via PPAR binding with resultant transcription and subsequent increased cell turnover, a phenomenon known classically as tumor promotion. Thus, retinoid actions cannot be limited to the biological effects historically or conventionally attributed to those of vitamin A (retinol and its esters), but must be viewed as a hormone subset of the larger steroid/thyroid/vitamin D_3 superfamily having both positive and negative regulatory roles across a great many physiologic activities, all part of a complex interconnected web. Even the definition of what or what does and what does not constitute a retinoid becomes problematic.

With the discovery that RXRs form heterodimers, the signaling pathways became even more complex. RXR-chicken upstream ovalbumin promotor transcription factor (COUP) heterodimers can function as potent repressors of gene expression (Kliewer et al., 1992c). Even heterodimers within the same RAR–RXR subgroup can function as either repressors or activators of gene expression depending on the particular response elements (text diagram adapted from Mangelsdorf, 1994):

RXR Heterodimers as Master Regulators

Positive	Negative
RXR------RXR	RXR------COUP
→1→	→1→
RXR------PPAR	RXR------RAR
→1→	→1→
RXR------RAR	
→2→	
RXR------VDR	
→3→	
RXR------TR	
→4→	
RXR------RAR	
→5→	

Left column: RXR homodimers or heterodimers that positively regulate gene expression through HREs composed of tandem repeat sequences. Right column: RXR heterodimers that regulate gene expression through HREs composed of tandem repeats spaced by a single nucleotide. Arrows indicate AGGTCA, or related, half-sites, and numbers are tandem repeat sequences spaced by 1, 2, 3, 4, and 5 nucleotides.

Studies with the PPAR-RXR heterodimer showed that both 9-*cis*-RA and PPARs result in synergistic induction of gene expression (Kliewer et al., 1992b). The nuclear receptor heterodimer can interact with either ligand alone or both ligands simultaneously, resulting in different levels of activation. RXR is unique in its ability to function as a homodimer. PPAR, VDR, TR, and RAR all require RXR heterodimerization to become fully functional receptor complexes. The properties of RXR are controlled by the ligands all-*trans*-RA and 9-*cis*-RA. For example, the presence of surplus 9-*cis*-RA promotes RXR homodimer formation, which activates target genes with RXR response elements (RXREs). By contrast, a surplus of RAR and low concentrations of 9-*cis*-RA in the cell promote formation of the RAR–RXR complex, which depends on both all-*trans*- and 9-*cis*-RAs in the cell, controlling activation or repression of target genes with RAR response elements (RAREs). Both the relative concentration of each receptor in the cell and the concentrations of the various ligands play a complex role in specific gene regulation and increase the diversity of the signaling pathways (Mangelsdorf, 1994). The following text diagram shows how ligand availability and receptor concentration can effect activation of different RA-responsive gene networks:

Two Retinoid Signaling Pathways Mediated by RXRs

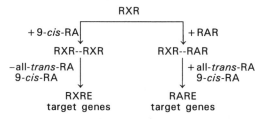

RXR can activate several retinoid-response gene networks depending on whether it complexes with itself or with RAR. Which complex is formed depends on ligand availability and receptor concentration. (Adapted from "Vitamin A Receptors," by D. J. Mangelsdorf, 1994, *Nutrition Reviews* 52(2):s32–s34.)

Therefore, not only do we have multiple actions of multiple receptors across the steroid/thyroid hormone superfamily controlling cell turnover and apparently unrelated physiologic functions, but local variations on the paradigm occur given the relative tissue-specific expression of receptor subtypes and concentrations of the cognate ligand.

Two research groups have independently generated null mutations in the mouse RXRα gene (Kastner et al., 1994; Sucov et al., 1994). Loss of function of the RXRα gene is lethal when bred to homozygosity. The major defect responsible for lethality is underdevelopment of the cardiac ventricle. Of note, ocular malformations were also observed. A phenotype synergy was observed when the RXRα mutation was introduced into RARα or RXRγ mutant backgrounds; there appears to be a functional convergence of the RXR and RAR signaling pathways in heart and eye morphogenesis. These are the first in vivo data from the level of the gene supporting the concept that RXR/RAR heterodimers act as transactivating transducers of RA signaling during development. It is noteworthy from the rat embryo data that combinations of excess all-*trans*-RA in the presence of excess thyroid hormone (T_3) produced a much greater than additive effect on rhombencephalic schisis, whereas 9-*cis*-RA plus T_3 produced a less than additive effect. Conversely, much greater than additive effects on anterior schisis were observed for 9-*cis*-RA plus T_3 whereas combined effects of all-*trans*-RA and T_3 were approximately additive (Creech Kraft et al., 1994c). Similarly, in *Xenopus* embryos, in which the brain, heart, and eyes are also the principal targets of RA teratogenicity (Durston et al., 1989; Sive et al., 1990; Creech Kraft et al., 1994a), recent studies have shown synergistic dysmorphogenic effects when embryos were exposed to a combination of 9-*cis*-RA and all-*trans*-RA during early neurulation (Creech Kraft & Juchau, 1995). Since not every human embryo exposed to Accutane has been born with malformations (Kassis et al., 1985), it is possible that mothers may be predisposed to dysmorphogenic effects in their embryos exposed to 13-*cis*-RA in the presence of high levels of T_3, due to undiagnosed mild metabolic disorders (hyperthyroidism), or abnormal levels of D_3, due to maternal deficiency or excessive intake of this vitamin, or to environmental exposure to PPARs, which are present in high-fat diets or in plasticizers and herbicides. Thus, environmental exposures to other ligands of retinoid-related receptors or to xenobiotics that alter retinoid metabolism may modify or otherwise predispose a particular pregnancy to retinoid teratogenic risk.

Yet another new family of retinoid-related "orphan" receptors termed the RZRs has been discovered, and in addition to receptor dimerization, retinoid receptor activity can be modulated by other nuclear transcription factors (e.g., AP-1). RZRα has been identified from human umbilical vein endothelial cells and RZRβ from rat brain (Carlberg et al., 1994). These RZR subtypes represent members of a new family of orphan receptors that most likely regulate specific gene expression. Sequence comparison reveals great similarity to the RAR and RXR receptors. The RZRs can function as both monomers and homomers. Northern blot analysis showed RZRβ messenger RNA only in the brain, whereas RZRα was expressed in many tissues. An appropriate ligand has not yet been identified for these receptors. Well-described ligands, such as T_3 and all-*trans*- and 9-*cis*-RA, do not augment RZR transactivation measured in their absence; however, transactivation can be enhanced by unidentified components of fetal calf serum. The RZRβ competes with the RAR/RXR heterodimer for activation of the cellular retinol binding protein (CRBP) response element. AP-1 is a complex of proto-oncoproteins *jun* and *fos*, whose activity can be stimulated by growth factors and tumor promotors. RA blocks the expression of the AP-1 complex at the transcriptional level, down-regulating transcriptional activation by the oncogenes *jun* and *fos* (Yang-Yen et al., 1990; Schuele et al., 1991). This interference is the result of protein–protein interactions between RARs and the AP-1 protein complex, leading to the mutual loss of the ability of either transcription factor to bind to DNA, that is, their own appropriate response elements. The RZRs and interactions with nuclear transcription factors again demonstrate the overwhelming, complex web of cellular control activities in which the retinoids are principal players.

BIOTRANSFORMATION OF VITAMIN A
TO ACTIVE METABOLITES

Isotretinoin (Accutane) is a naturally occurring metabolite of retinol (vitamin A_1), and it is the trade name for 13-*cis*-RA. After oral administration of 13-*cis*-RA to humans at therapeutic doses, four metabolites were identified in the serum (Creech Kraft et al., 1991b), including 4-oxo-all-*trans*-RA, 4-oxo-13-*cis*-RA, all-*trans*-retinoyl-β-glucuronide (all-*trans*-RAG) and all-*trans*-RA (Figure 2-8). High levels of all-*trans*-RA were detected in an embryo carried by a woman who had been therapeutically treated with Accutane (Creech Kraft et al., 1989a). The levels of both all-*trans*-RA and 13-*cis*-RA in the abortus were 10-fold higher than normal average or endogenous RA levels, whereas levels of all-*trans*-retinol were in a normal range (Creech Kraft et al., 1993). Two independent laboratories have also verified the presence of all-*trans*-3,4-didehydroretinol (all-*trans*-dd-retinol) in human embryos during embryogenesis (Creech Kraft & Juchau, 1994; Sass, 1994).

Possibly the teratogenic effects of 13-*cis*-RA in humans are due to metabolic conversion to all-*trans*-RA. The following discoveries strengthen this idea: 13-*cis*-RA is a potent human teratogen, but is only marginally teratogenic in the mouse at doses 100 times higher than those used in human therapy; the all-*trans* isomer, on the other hand is a potent teratogen in the mouse at relatively low doses (Kochhar et al., 1984). Transplacental pharmacokinetics showed that all-*trans*-RA accumulated much more extensively than its *cis* isomer in mouse embryos (Creech Kraft et al., 1987, 1989b; Satre et al., 1989). In those studies employing single oral administrations of all-*trans*-RA or 13-*cis*-RA, the areas under the concentration–time curves in embryonic tissues were 30-fold higher

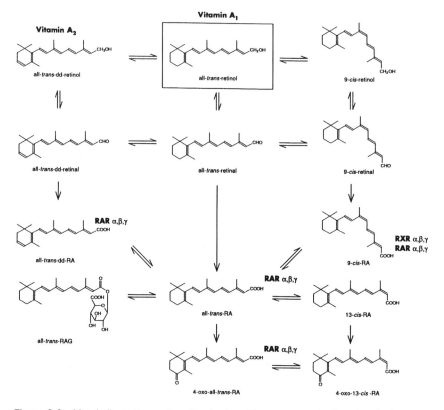

Figure 2-8 Metabolic pathway for vitamin A$_1$ with cross-over to the vitamin A$_2$ metabolic pathways. The retinoids that are high-affinity ligands for the nuclear receptors (RARs and RXRs) are indicated.

for the *trans* isomer than for the *cis* isomer. The teratogenicity of 13-*cis*-RA observed after multiple high oral dosing to pregnant mice was hypothetically due to its conversion to the all-*trans* isomer. The *trans* isomer was subsequently detected at high concentrations in the mouse embryo for 10 critical hours during gestation (Creech Kraft et al., 1991a). 4-Oxo-all-*trans*-RA is a potent teratogen in most species (Willhite et al., 1989; Pijnappel et al., 1993), whereas in mice 4-oxo-13-*cis*-RA is only marginally teratogenic (Kochhar & Penner, 1987; Satre et al., 1989; Creech Kraft et al., 1989b). Embryonic exposure to 4-oxo-all-*trans*-RA was far higher than to its corresponding *cis* isomer, because only the *trans* isomer accumulated in the mouse embryo (Creech Kraft et al., 1989b). Gunning et al. (1993) reported that the glucuronide conjugate of RA, all-*trans*-RAG, was not teratogenic when administered orally to rats during gestation. Two independent research groups have concluded that the teratogenic effects of retinol (Morriss, 1972) are due to its conversion to all-*trans*-RA (Kochhar et al., 1988; Eckhoff et al., 1989). Comprehensive studies using intraamniotic microinjections in rat whole embryos revealed that all-*trans*-RA and 4-oxo-all-*trans*-RA were far more potent than their corresponding *cis* isomers and precursors, retinol or all-*trans*-RAG (M.-O. Lee et al., 1991; Creech Kraft & Juchau, 1992a, 1992b; Creech Kraft et al., 1992a, 1992b). High-performance liquid chromatography (HPLC) analyses indicated that biotransformation to all-*trans*-RA and 4-oxo-all-*trans*-RA by the conceptus appeared responsible for the dysmorphogenic effects produced by intraamniotic microinjections of the parent polyene retinoids.

In contrast to the findings in mice, several investigations in rats, rabbits, and monkeys have employed a repetitive dosing regimen during organogenesis (Collins et al., 1994; Sandberg et al., 1994; Eckhoff et al., 1994). These authors conclude that both 13-*cis*-RA and retinol have intrinsic teratogenic activity without a required biotransformation to all-*trans*-RA. All the retinoid pharmacokinetic studies listed here were conducted during fetal life (following the sixth daily administration) and therefore did not reflect embryonic exposure to the parent retinoid and its metabolites at earlier, more sensitive stages. These data also did not reflect the important observation that repeated administration of all-*trans*-RA induces its own metabolism, whereas 13-*cis*-RA does not (Kalin et al., 1984; Creech Kraft et al., 1991a, 1991b); studies employing repetitive dosing of these two polyene retinoids, in particular, may result in different findings. Collins et al. (1995, p. 9) found that "multiple administration of a drug such as RA, which induces its own elimination pathways, results in substantially lowered drug levels in maternal plasma and embryo." Because of metabolic considerations, single doses administered at specific developmental periods are expected to be more effective in uncovering particular developmental effects (Collins et al., 1995). Repetitive dosing is useful in the human health risk assessment and in calculation of therapeutic index between different retinoids, but single-exposure studies appear to have greatest utility in laboratory screening for teratogenic activity and in mechanism of action analyses.

All-*trans*-RA, 4-oxo-all-*trans*-RA, and all-*trans*-dd-RA bind and activate retinoic acid receptors α, β, and γ (RARs α, β, and γ) (Petkovich et al., 1987; Brand et al., 1988; Krust et al., 1989), whereas their corresponding 13-*cis* isomers do not (Apfel et al., 1992; Allenby et al., 1993; Alam et al., 1995). If 13-*cis*-RA is indeed an intrinsic dysmorphogen, it would be of interest to discover its mechanism of action because 13-*cis*-RA does not bind or activate the presently known nuclear receptors. Soprano et al. (1994) reported that there is a sustained elevation in RARβ2 mRNA and protein during RA-induced fetal dysmorphogenesis. Similarly, Willhite et al. (1996) suggest from their recent studies with synthetic RAR-selective ligands that inappropriate or prolonged RAR transactivation is linked with retinoid embryopathy. 9-*cis*-RA binds and activates the RAR and RXR nuclear receptors (Heyman et al., 1992; Levin et al., 1992). After oral administration to dams and in some in vitro assays, 9-*cis*-RA and RXR-selective ligands are generally of much lower teratogenic potential than RAR-selective ligands (Collins et al., 1993; Kochhar et al., 1993; Dawson et al., 1993). Ratios of embryonic to maternal plasma concentrations suggest a limited placental transfer of 9-*cis*-RA to mouse and rat embryos (Tzimas et al., 1994). However, the observations that 9-*cis*-RA is an equipotent dysmorphogen in cultured rat embryos compared to all-*trans*-RA, but causes higher incidence of missing optic vesicles and cardiac defects (Creech Kraft & Juchau, 1993), supports the idea that retinoids that bind and activate nuclear receptors can be considered direct-acting teratogens, given the appropriate dose and stage of development. Using *Xenopus* embryos in which the spatial and temporal distribution of RAs and their precursors has been established (Creech Kraft et al., 1994a, 1994b), it has been shown that after retinoid treatment during early neurulation, only the receptor ligands all-*trans*-RA, 9-*cis*-RA, all-*trans*-dd-RA, and 4-oxo-all-*trans*-RA are direct-acting dysmorphogens, all causing cardiac, eye, and brain malformations. All of these ligands bind and transactivate the RARs, but not all bind and transactivate the RXRs. This is further verification of the hypothesis that retinoid dysmorphogenesis is mediated through retinoid nuclear receptor binding and transactivation of downstream genes and points toward the RAR subfamily as most directly involved in retinoid embryopathy.

SYNTHETIC RETINOIDS: BINDING AND ACTIVATION OF NUCLEAR RECEPTORS

There are a number of features in the preceding working hypothesis that remain unexplained. For example, in vitro measures of retinoid binding and RAR transactivation do not necessarily correlate with retinoid teratogenic potency. Binding to wild-type RARα, β, and γ (measured from competition with all-*trans*-RA) and transcriptional transactivation activity (as measures in monkey kidney CV-1 cells) for all-*trans*-RA, SRI 5898-71 (TTNN), Ro 13-7410 (TTNPB), and SR 3961 (TTAB) found binding for all-*trans*-RA, TTNPB, and TTAB of generally the same order of magnitude (measured as the concentration of retinoid necessary to displace 50% of [^{3}H]all-*trans*-RA from the receptor, IC50). The binding of TTNN to RARs was only 2–32% that of all-*trans*-RA. The transactivation EC50 values for TTNN were equal to (RARγ), less than (RARβ), or greater than (RARα) those for all-*trans*-RA. Transactivation EC50 values for TTNPB were always at least one order of magnitude less than those for all-*trans*-RA and for TTAB were up to 60 times less than those for all-*trans*-RA (Dawson et al., 1993). TTNN, TTNPB, and TTAB teratogenic potency in hamster was 34, 750–1000, and 8000 times that of all-*trans*-RA, respectively (Willhite & Dawson, 1990; Willhite et al., 1996). While elimination of such retinoids can be much slower than for polyene congeners (Howard et al., 1989), neither their pharmacokinetic parameters nor their CRABP or nuclear receptor binding and transactivation profiles can account for the marked potency of these conformationally restricted retinoids. RARβ2 mRNA levels in mouse embryos were increased after TTNPB (Ro13-7410) treatment (just as was the case after a teratogenic dose of all-*trans*-RA or 13-*cis*-RA; Soprano et al., 1994), but the magnitude of

the increase after TTNPB exposure was no greater than after exposure to teratogenic polyene retinoids (Jiang et al., 1994). Prolonged retention of these metabolically stable congeners (as compared to the relatively rapid oxidation and
conjugation of retinoic acids) could account for prolonged receptor binding and
transactivation, but only extremely small amounts of these compounds cross the
placenta (even after a teratogenic dose) (Howard et al., 1989), and only very
small doses (e.g., 750 ng/kg as a single TTAB oral intubation) can be tolerated
by the dam. Quantitative disposition data for these compounds are largely non-
existent, but available metabolism studies point to unidentified biotransformation products (Pignatello et al., 1995), some of which may possess increased
receptor affinity or increased transactivation efficiency.

In general retinoid teratogenic potency in vivo tends to parallel retinoid
teratogenesis as measured in cultured mammalian whole embryos and limb buds.
Retinoids having activity in these systems tend also to show the greatest prom-
ise in preclinical screens for chemopreventive and dermatologic applications.
This general conclusion has as its basis the fact that it is the retinoid nuclear
receptors expressed during embryogenesis and in preneoplastic, neoplastic, or
otherwise abnormal cells that are the common molecular targets. The corollary
then to this conclusion is that a fundamental or complete separation of retinoid
chemopreventive/chemotherapeutic actions from developmental toxicity cannot
in all likelihood be realized, particularly with congeners having measurable RAR
affinity. Nevertheless, identification of retinoids with substantially improved
therapeutic ratios, combined with routes of administration (topical or otherwise
targeted) with therapy restricted to males, postmenopausal females, or females
in compliance with current FDA guidelines for isotretinoin therapy, should mitigate this obstacle to the diverse therapeutic benefits of these drugs.

RETINOIC ACID AND DEVELOPMENT

There is circumstantial evidence for a vitamin A requirement in normal human
embryonic development (Willhite et al., 1989). This requirement has been demonstrated in farm and laboratory animals (pigs, rabbits, rats, mice), and it has
been known for more than 60 years (Hale, 1933). Perhaps the clearest demonstration of retinoid participation in normal embryogenesis comes from data gathered
in chick and frog (for review, see Maden, 1994a; Hofmann & Eichele, 1994).
These animals have several advantages over mammals because chick and frog
embryos are easily obtained, are available in large numbers, are inexpensive,
and are easily manipulated. *Xenopus* embryos provide a suitable model for the
study of embryonic retinoid metabolism during early neurulation (Creech Kraft
et al., 1995a, 1995b; Creech Kraft & Juchau, 1995). Avian and amphibian systems are at a disadvantage in the human health risk assessment arena, of course,
in that maternal metabolism and the role of the placenta are ignored, but for
direct studies chick and frog simplify data collection, experimental design,
and interpretation. The chick and frog yield insight into retinoid control of

embryonic pattern execution that would be extremely difficult to obtain otherwise.

Chick wing pattern formation has been the most intensively studied system in regard to RA control of patterning. All-*trans*-RA treatment can act to alter specific patterns of cell differentiation leading to changes in positional values, such that anterior cells behave as though they were located in posterior areas. Regardless of whether cells from the chick limb bud posterior margin (the polarizing region or zone of polarizing activity, ZPA) are transplanted to the anterior margin or whether those chick anterior cells are treated with all-*trans*-RA, the consequence is identical—mirror image duplications (Tickle & Brickell, 1991). This activity is consistent across species; when mouse limb bud cells having polarizing activity are transplanted to the anterior aspect of the chick limb, the response in the chick wing is the same. There is a clear retinoid concentration gradient with higher (1.4–2.5 times) concentrations of all-*trans*-RA in posterior cells as compared to anterior cells (Thaller & Eichele, 1987), but concentrations of its metabolic precursor, all-*trans*-retinol, are uniform from anterior to posterior. Retinol has only very limited ability to influence limb bud cell positional values, implying the existence of an autocatalytic mechanism wherein up-regulation of the nuclear retinoic acid receptor (RARβ) gene and the CRBP I gene then act in concert with CRABPs to modulate and control local production of all-*trans*-RA from retinol (Summerbell & Waterson, 1993). There is a sophisticated interplay in local control of retinoid metabolism, nuclear receptor transactivation, and a cascade of gene activation and translation that depends on the concentrations of those free retinoids that are capable of receptor binding and transactivation; in turn, the cellular retinoid binding proteins affect the free concentrations of ligands and act to maintain normal homeostasis (Ross, 1993).

The principal retinoid found in chick wing bud is not all-*trans*-RA, but all-*trans*-dd-RA, which arises from local oxidation of all-*trans*-dd-retinol (known as vitamin A$_2$) (Thaller & Eichele, 1990) (Figure 2-8). The didehydro-RA analog, like all-*trans*-RA, is a ligand for the RARs. This acidic congener binds CRABPs, and it is equipotent to all-*trans*-RA in producing mirror-image duplications in chick wing. Unlike all-*trans*-RA, which has a somewhat higher concentration in posterior sections as compared to anterior, all-*trans*-retinol, all-*trans*-dd-retinol, and all-*trans*-dd-RA are uniform across the wing bud (Scott et al., 1994). All-*trans*-RA is present in the peripheral mesenchyme of the limb bud and (based on the distribution of retinol, its cellular binding protein, and the CRABPs) it has been suggested that all-*trans*-RA is produced normally in ectoderm and then transferred to the peripheral mesenchyme (Gustafson et al., 1993). While the precise roles of all-*trans*-RA and all-*trans*-dd-RA in limb development are not known, the data published to date suggest that all-*trans*-RA is most closely tied to induction of the expression of *sonic hedgehog* gene morphogenetic activity, which participates in normal pattern determination through positional signaling and digit specification (Riddle et al., 1993). Current information suggests the didehydro analogue is involved with normal programmed cell death and necrosis of whole digits (in chick) or interdigital mesenchyme (in tri- and pentadactyl animals) (Scott et al., 1994).

The findings in chick, by and large, can be carried over to amphibians and mammals. Retinoids can elicit reduction and abnormal differentiation in amphibian limb and fin along the proximodistal axis (Niazi & Saxena, 1968). Retinoids also alter positional values in blastema along anteroposterior (A/P) and dorsoventral axes (Durston et al., 1989), with the latter leading to anterior truncation. In amphibia, the dorsal ectoderm is a specific target of RA action, where RA inhibition of anterior specific ectodermal gene expression (neural and cement gland genes at least) is dependent on intermediate protein synthesis—implying that while RA itself and, by extension, the RARs are not direct mediators of those genes, the intermediate products of RA/RAR-controlled genes are potential second messengers (Sive et al., 1990). A/P axis formation occurs during gastrulation, and at the same time the polarity of the dorsal mesoderm becomes evident; the dorsal ectoderm then appears to acquire from the mesoderm its particular fate where it gives rise to neural (brain, spinal cord) and nonneural (cement gland, fin) tissues. RA can induce 10-fold elevations in the expression of frog homeodomain genes in the dorsal ectoderm, but ectoderm from RA-treated embryos cannot respond normally to transplanted mesoderm (Sive & Chen, 1991). RA has a role in determination of the normal lineage of mesodermal cells and controls the inducing properties of the anterior mesoderm (Sive & Chen, 1991). RA also has direct actions in the neural ectoderm and can elicit abnormal expression of *Hox* gene clusters normally expressed in the anterior (fore)brain. RA appears to be involved in normal hindbrain patterning by regulating the A/P domains of homeobox gene expression whose normal boundaries correspond to the anatomical boundaries of the rhombomeres. Both 9-*cis*-RA (Creech Kraft et al., 1993) and all-*trans*-RA (Durston et al., 1989; Creech Kraft et al., 1993) are endogenously present in both the dorsal anterior and posterior regions of *Xenopus* embryos during early development, as are RARγ (Ellinger-Ziegelbauer & Dreyer, 1991; Sharpe, 1992), RXR (Blumberg et al., 1992; Marklew et al., 1994), and CRABP (Dekker et al., 1994; Ho et al., 1994). It is likely that RA modulation of intermediate players, like the polypeptide growth factors (Kimelman & Kirschner, 1987; Tabin, 1991; Kimelman et al., 1992; Mahmood et al., 1993), regulates the character of the axial mesoderm through modifying ectodermal response to mesoderm-derived messengers (Ruiz i Altaba & Jessell, 1991).

RETINOIC ACID AND GENE EXPRESSION

It has long been known that excess RA causes craniofacial malformations in both birds and mammals where the CNS, the chondrocranium and viscerocranium are affected (Lammer et al., 1985; Sulik et al., 1988). The hindbrain and the neural crest cells derived from it are sensitive to RA (Morriss, 1972; Morriss & Thorogood, 1978) in that RA is capable of changing the segmental organization of the developing hindbrain. It is currently believed that these changes are in part due to RA's ability to alter the pattern of expression of genes involved in the specification of cell identity. Among these genes are segment polarity genes, such as *sonic hedgehog*; homeobox genes, such as *engrailed* and *hox* genes; and zinc-finger-containing DNA-binding proteins, such as RARβ and *Krox 20* (Sucov et al., 1990; de The et al., 1990; Morriss-Kay et al., 1991; Holder & Hill, 1991; Conlon & Rossant, 1992; Marshall et al.,1992). If RA levels are raised prior to segmentation of the hindbrain into rhombomeres, the expression of these genes may be shifted in position, increased in size (*Hox*) (Morriss-Kay et al., 1991; Kessel & Gruss, 1991), or even lost completely (*Krox 20*) (Papalopulu et al., 1991). The normal function of RA can best be understood in terms of control of gene expression during embryogenesis.

Homeobox genes contain segments of DNA arranged through a helix-turn-helix motif called the homeobox. They were first discovered in *Drosophila* and amazingly, these genes were lined up on the chromosome in exactly the same order as the anterior–posterior sequence of the segments of the fly's body they controlled. This short sequence of DNA, the homeobox, is identical in all homeobox genes across all invertebrate and vertebrate species. Numerous studies have demonstrated that the products of homeobox genes regulate the activity of other genes. One notable class of vertebrate homeobox genes are the *Hox* genes. Higher vertebrates, including humans, have four clusters of *Hox* genes, with each cluster on a different chromosome, and these genes are normally expressed during vertebrate development. In limb buds (Davidson et al., 1991) and along the anteroposterior body axis, their expression seems to follow a spatial colinearity rule (Graham et al., 1991). Along the anteroposterior body axis, 3' genes in each of the 4 Hox clusters are expressed earlier in more anterior regions of the embryo, while 5' genes are expressed more posteriorly, and later in development. Some of these clusters of homeobox genes are turned on in a sequential manner by treatment with RA (Simone et al., 1990). Thus, RA is thought to convey positional information by controlling local segmented gene expression (Rossant et al., 1991; Conlon & Rossant, 1992; Tabin, 1991). This same retinoic acid–*Hox* gene paradigm appears to hold across all mammals, including human beings (Stornaivolo et al., 1990).

A segment polarity gene *sonic hedgehog*, which is related to the *Drosophila* segment polarity gene, *hedgehog* (Mohler & Vani, 1992; J. Lee et al., 1992; Tabata et al., 1992), has been identified in mouse (Echelard et al., 1993), chicken (Riddle et al., 1993), and frog (Ekker et al., 1995). *Sonic hedgehog* is expressed in the notochord and in the ZPA, both of which are signaling centers that are thought to mediate the central nervous system and limb bud polarity, respectively. *Sonic hedgehog* encodes a secreted factor that is produced by the ZPA, and all-*trans*-RA activates *sonic hedgehog* activity (Riddle et al., 1993), leading to the expression of *Hox* genes.

The vertebrate organizer was originally described when it was found that the dorsal lip of *Xenopus* embryos could change the fate of surrounding cells and cause formation of a secondary axis when it was transplanted into the ventral region of a host embryo. Due to the inductive properties of the dorsal blastopore lip, it was named the organizer (Spemann & Mangold, 1924). The *Xenopus* homeobox gene *Xlim-1* is also expressed in the organizer region in the anterior dorsal mesoderm that underlies the head region (Taira et al., 1992, 1994a, 1994b). The mouse homologue of *Xlim-1* was cloned (Fujii et al., 1994; Barnes et al., 1994), and *Lim-1* null mice were generated by gene targeting in embryonic stem cells (Shawlot & Behringer, 1995). These mice lacked anterior head structures, but the remaining body axis developed normally. Therefore the homeobox *Lim-1* type genes are required for formation of the early organizer node and they convey positional information to the anterior axial mesoderm during mouse embryogenesis. It has been shown that both all-*trans*-RA and 9-*cis*-RA induce this gene in *Xenopus* animal caps (Taira et al., 1992; Creech Kraft et al., 1994a) and are endogenously available in the organizer region, giving evidence that this gene may be under the normal control of RAs and their receptors.

The extracellular matrix is critical for cellular migration during morphogenesis, including migration of the neural crest. There are numerous other genes that are involved in the transcriptional control of matrix proteins that are regulated by RA. These include collagens, fibronectins, and basement membrane proteins (for review see Gudas et al., 1994). Likewise, programmed cell death (apoptosis), which has been well described during normal and abnormal development of the nervous system (Sulik et al., 1988), limb (Davies et al., 1992), and palate, is believed to be controlled by RA-inducible genes. One example is type II transglutaminase, an enzyme that is both induced and activated by RA and causes intracellular protein cross-linking, which may well be required for apoptotic cell death (Davies et al., 1992).

CONCLUSIONS

While there is no a priori reason to believe that retinoid embryopathy results from disturbance of the same mechanism by which retinoids participate in normal embryogenesis, understanding of a requirement for RA in normal embryogenesis can help explain the marked morphologic changes observed in retinoid-induced teratogenesis. Embryonic development involves a tightly controlled series of changes in gene expression in time and space, and it is now well established that pharmacologic exposure of vertebrate embryos to retinoids results in inappropriate patterns of gene expression (Morriss-Kay, 1993; Gudas, 1994; Gudas et al., 1994). There is circumstantial evidence that it is the abnormal and perhaps inappropriately prolonged transcriptional transactivation by retinoid nuclear receptors of critical genes in the target embryonic cells that forms the basis for retinoid teratogenesis. This hypothesis stems from the following observations. First, all retinoids that have teratogenic actions in vitro or in vivo are or can be transformed to a compound with an acidic polar terminus. Second, among naturally occurring retinoids, only those that exist as free acid are capable of receptor binding and transactivation. Third and importantly, the retinoid nuclear receptors are expressed at the appropriate times and in those embryonic cells affected by elevated retinoids.

REFERENCES

Adams, J. 1993. Structure-activity and dose-response relationship in the neural and behavioral teratogenesis of retinoids. *Neurotoxicol. Teratol.* 15:193–202.

Adams, J., and Lammer, E. J. 1991. Relationship between dysmorphology and neuro-psychological function in children exposed to isotretinoin "in utero." In *Functional neuroteratology of short-term exposure to drugs*, eds. T. Fuji and G. J. Boer, pp. 159–170. Tokyo: Tokyo University Press.

Alam, M., Zhestkov, V., Sani, B. P., Venepally, P., Levin, A. A., Kazmer, E. L., Norris, A. W., Zhang, X.-K., Lee, M.-O., Hill, D., Lin, T. S., Brouilette, W., and Muccio, D. 1995. Conformationally defined 6-s-*trans*-retinoic acid analogs. 2. Selective agonists for nuclear receptor binding and transcriptional activity. *J. Med. Chem.* 38:2302–2310.

Allenby, G., Bocquel, M.-T., Saunders, M., Kazmer, S., Speck, J., Rosenberger, M., Lovey, A., Kastner, P., Grippo, J., Chambon, P., and Levin, A. 1993. Retinoic acid receptors and retinoid X receptors: Interactions with endogenous retinoic acids. *Proc. Natl. Acad. Sci. USA* 90:30–34.

Apfel, C., Crettaz, M., and LeMotte, P. 1992. Differential binding and activation of synthetic retinoids to retinoic acid receptors. In *Retinoids in normal development and teratogenesis*, ed. G. Morriss-Kay, pp. 65–74. Oxford: Oxford University Press.

Astrom, A., Tavakkol, A., Pettersen, U., Cronmie, M., Elder, J. T., and Voorhees, J. J. 1991. Molecular cloning of two human cellular retinoic acid-binding proteins (CRABP). Retinoic acid-induced expression of CRABP-II but not CRAPB-I in adult human skin *in vivo* and in skin fibroblasts *in vitro*. *J. Biol. Chem.* 266:17662–17666.

Barnes, J. D., Crosby, J. L., Jones, C. M., Wright, C. V. E., and Hogan, B. L. M. 1994. Embryonic expression of *Lim-1*, the mouse homolog of *Xenopus Xlim-1* suggests a role in lateral mesoderm differentiation and neurogenesis. *Dev. Biol.* 161:168–178.

Beato, M. 1989. Gene regulation by steroid hormones. *Cell* 56:335–344.

Bellairs, R., Ede, D. A., and Lash, J. W., eds. 1986. *Somites in developing embryos*. New York: Plenum Press.

Blumberg, B., Mangelsdorf, D. J., Dyck, J. A., Bittner, D. A., Evans, R. M., and De Robertis, E. M. 1992. Multiple retinoid-responsive receptors in a single cell: Families of retinoid X receptors and retinoic acid receptors in the *Xenopus* egg. *Proc. Natl. Acad. Sci. USA* 89:2321–2325.

Boylan, J. F., and Gudas, L. J. 1992. The level of CRABPI expression influences the amounts and types of all-*trans*-retinoic acid metabolites in F9 teratocarcinoma stem cells. *J. Biol. Chem.* 267:21486–21491.

Brand, N. J., Petkovich, M., Krust, A., Chambon, P., de The, H., Machio, A., Tiollais, P., and Dejean, A. 1988. Identification of a second human retinoic acid receptor. *Nature* 332:850–853.

Braun, J. T., Franciosi, R. A., Mastri, A. R., Drake, R. M., and O'Niel, B. L. 1984. Isotretinoin dysmorphic syndrome. *Lancet* 1:506–507.

Burk, D. T., and Willhite, C. C. 1992. Inner ear malformations induced by isotretinoin in hamster fetuses. *Teratology* 46:147–157.

Carlberg, C., van Huijsduijnen, R. H., Straple, J. R., DeLamarter, J. F., and Becker-Andre, M. 1994. RZRs, a new family of retinoid related orphan receptors that function as monomers and homodimers. *Mol. Endocrinol.* 8:757–770.

Chytil, F., and Ong, D. 1978. Cellular Vitamin A binding proteins. *Vitam. Horm.* 36:1–32.

Collins, M. D., Schreiner, C. M., and Scott, W. J. 1993. Comparative potency of all-*trans*- and 9-*cis*-retinoic acid for the induction of murine exencephaly. *Teratology* 47:385.

Collins, M., Tzimas, G., Hummler, H., Buergin, H., and Nau, H. 1994. Comparative teratology and transplacental pharmacokinetics of all-*trans*-retinoic acid, 13-*cis*-retinoic acid and retinyl palmitate following daily administration in rats. *Toxicol. Appl. Pharmacol.* 127:132–144.

Collins, M., Tzimas, G., Burgin, H., Hummler, H., and Nau, H. 1995. Single versus multiple dose administration of all-*trans*-retinoic acid during organogenesis: Differential metabolism and transplacental kinetics in rat and rabbit. *Toxicol. Appl. Pharmacol.* 130:9–18.

Conlon, R. A., and Rossant, J. 1992. Exogenous retinoic acid rapidly induces anterior expression of murine *Hox-2* genes *in vivo*. *Development* 116:357–368.

Creech Kraft, J., and Juchau, M. R. 1992a. Correlations between conceptual concentrations of all-*trans*-retinoic acid and dysmorphogenesis after microinjections of all-*trans*-retinoic acid, 13-*cis*-retinoic acid, all-*trans*-retinoyl-β-glucuronide or retinol in cultured whole rat embryos. *Drug Metab. Dispos.* 20:218–225.

Creech Kraft, J., and Juchau, M. R. 1992b. Conceptual biotransformation of 4-oxo-all-*trans*-retinoic acid, 4-oxo-13-*cis* retinoic acid and all-*trans*-retinoyl-β-glucuronide in rat whole embryo culture. *Biochem. Pharmacol.* 43:2289–2292.

Creech Kraft, J., and Juchau, M. R. 1993. 9-*cis*-Retinoic acid: A direct-acting dysmorphogen. *Biochem. Pharmacol.* 43:2289–2292.

Creech Kraft, J., and Juchau, M. R. 1994. 3,4-Didehydroretinol may be present in human embryos/fetuses. *Reprod. Toxicol.* 8:191.

Creech Kraft, J., and Juchau, M. R. 1995. *Xenopus laevis*: A model system for the study of embryonic retinoid metabolism. III. Isomerization and metabolism of all-*trans*-retinoic acid and 9-*cis*-retinoic acid and their dysmorphogenic effects in embryos during neurulation. *Drug Metab. Dispos.* 23:1058–1072.

Creech Kraft, J., Kochhar, D. M., Scott, W. J., Jr., and Nau, H. F. 1987. Low teratogenicity of 13-*cis*-retinoic acid (isotretinoin) in the mouse corresponds to low embryo concentrations during organogenesis: Comparison to the all-*trans* isomer. *Toxicol. Appl Pharmacol.* 87:474–482.

Creech Kraft, J., Nau, H., Lammer, E., and Olney, A. 1989a. Embryonic retinoid concentrations after maternal intake of isotretinoin. *N. Engl. J. Med.* 321:262.

Creech Kraft, J., Lofberg, B., Chahoud, I., Bochert, G., and Nau, H. 1989b. Teratogenicity and placental transfer of all-*trans*, 13-*cis*, 4-oxo-all-*trans*, and 4-oxo-13-*cis*-retinoic acid after a low oral dose during organogenesis in mice. *Toxicol. Appl. Pharmacol.* 100:162–176.

Creech Kraft, J., Eckhoff, C., Kochhar, D. M., Bochert, G., Chahoud, I., and Nau, H. 1991a. Isotretinoin (13-*cis*-retinoic acid), metabolism, cis-*trans* isomerization, glucuronidation and transfer to the mouse embryo: Consequences for teratogenicity. *Teratogen. Carcinogen. Mutagen.* 11:21–30.

Creech Kraft, J., Slikker, W., Jr., Bailey, J. R., Roberts, L. G., Fischer, B., Witffoht, W., and Nau, H. 1991b. Plasma pharmacokinetics of 13-*cis* and all-*trans* retinoic acid in the cynomolgus monkey and the identification of the conjugate metabolites 13-*cis* and all-*trans*-retinoyl-β-glucuronides: A comparison to one human case study with isotretinoin. *Drug Metab. Dispos.* 19:317–324.

Creech Kraft, J., Bechter, R., Lee, Q. P., and Juchau, M. R. 1992a. Microinjections of cultured rat conceptuses: Studies with 4-oxo-all-*trans*-retinoic acid, 4-oxo-13-*cis*-retinoic acid and all-*trans*-retinoyl-β-glucuronide. *Teratology* 45:259–270.

Creech Kraft, J., Bui, T., and Juchau, M. R. 1992b. Elevated levels of all-*trans*-retinoic acid in cultured rat embryos 1.5 hr after microinjections with 13-*cis*-retinoic acid or retinol and correlations with dysmorphogenesis. *Biochem. Pharmacol.* 44:21–24.

Creech Kraft, J., Shepard, T., and Juchau, M. R. 1993. Tissue levels of retinoids in human embryo/fetuses unexposed to Accutane[R]. *Reprod. Toxicol.* 7:11–15.

Creech Kraft, J., Schuh, T., Juchau, M. R., and Kimelman, D. 1994a. The retinoid X receptor, 9-*cis*-retinoic acid, is a potential regulator of early *Xenopus* development. *Proc. Natl. Acad. Sci. USA* 91:3067–3071.

Creech Kraft, J., Schuh, T., Juchau, J., and Kimelman, D. 1994b. Temporal distribution, localization and metabolism of retinol, didehydroretinol and retinal during development. *Biochem. J.* 301:111–119.

Creech Kraft, J., Willhite, C. C., and Juchau, M. R. 1994c. Embryogenesis in cultured whole rat embryos after combined exposure to 3,3',5-triiodo-L-thyronine (T3) plus all-*trans*-retinoic acid and to T3 plus 9-*cis* retinoic acid. *J. Craniofac. Genet. Dev. Biol.* 14:75–86.

Creech Kraft, J., Kimelman, D., and Juchau, M. R. 1995a. *Xenopus laevis*: A model system for the study of retinoid metabolism 1. Embryonic metabolism of 9-*cis*- and all-*trans*-retinals and retinols to their corresponding acid forms. *Drug Metab. Dispos.* 23:72–82.

Creech Kraft, J., Kimelman, D., and Juchau, M. R. 1995b. *Xenopus laevis*: A model system for the study of retinoid metabolism 2. Embryonic metabolism of all-*trans*-3,4-didehydroretinol to all-*trans*-3,4-didehydroretinoic acid. *Drug Metab. Dispos.* 23:83–89.

Davidson, D. R., Crawley, A., Hill, R. E., and Tickle, C. 1991. Position dependent expression of two related homeobox genes in developing vertebrate limbs. *Nature* 352:429–431.

Davies, P. J. A., Chiocca, E. A., Basilon, J. P., Gentile, V., Thomazy, V., and Fesus, L. 1992. Retinoid-regulated expression of transglutaminase links to the biochemistry of programmed cell death. In *Retinoids in normal development and teratogenesis*, ed. G. Morriss-Kay, pp. 249–263. Oxford: Oxford University Press.

Dawson, M. I., Hobbs, P. D., Stein, R. B., Beyer, T. S., and Heyman, R. A. 1993. Interaction of retinoids with retinoic acid nuclear receptor isoforms. In *Retinoids. Progress in research and clinical applications*, eds. M. A. Livrea and L. Packer, pp. 205–221. New York: Marcel Dekker.

Dekker, E.-J., Vaessen, M.-J., van den Berg, C., Timmermans, A., Godsave, S., Holling, T., Nieuwkoop, P., van Kessel, A., and Durston, A. 1994. Overexpression of a cellular retinoic acid binding protein (xCRABP) causes anteroposterior defects in developing *Xenopus* embryos. *Development* 120:973–985.

Dencker, L., Gustafson, A. L., Annerwall, E., Busch, C., and Eriksson, U. 1991. Retinoid-binding proteins in craniofacial development. *J. Craniofac. Genet. Dev. Biol.* 11:303–314.

de The, H., Vivanco-Ruiz, M. M., Tiollais, P., Stunnenberg, H., and Dejean, A. 1990. Identification of a retinoic acid response element in the retinoic acid receptor β gene. *Nature* 343:177–180.

Deuster, G., Shean, M. L., McBride, M. S., and Stewart, M. J. 1991. Retinoic acid response element in the human alcohol dehydrogenase gene ADH3: Implications for regulation of retinoic acid synthesis. *Mol. Cell Biol.* 11(3):1638–1646.

Dreyer, C., Krey, G., Keller, H., Givel, F., Helftenbein, G., and Wahli, W. 1992. Control of the peroxisome β-oxidation by a novel family of nuclear hormone receptors. *Cell* 68:879–887.

Durand, B., Saunders, M., Leroy, P., Leid, M., and Chambon, P. 1992. All-*trans*- and 9-*cis*-retinoic acid induction of CRABPII transcription is mediated by RAR-RXR heterodimers bound to DR1 and DR2 repeated motifs. *Cell* 71:73–85.

Durston, A. J., Timmermans, W. J., Hage, H. F. J., deVries, N. J., Heideveld, M., and Nieuwkoop, P. D. 1989. Retinoic acid causes an anteroposterior transformation in the developing central nervous system. *Nature* 340:140–144.

Echelard, Y., Epstein, D. J., St-Jacques, B., Shen, L., Mohler, J., McMahon, J. A., and McMahon, A. P. 1993. Sonic hedgehog, a member of a family of putative signaling molecules, is implicated in the regulation of CNS polarity. *Cell* 75:1417–1430.

Eckhoff, C., Loefberg, B., Chahoud, G., Bochert, G., and Nau, H. 1989. Transplacental pharmacokinetics and teratogenicity of a single dose of retinol (vitamin A) during organogenesis in the mouse. *Toxicol. Lett.* 48:171–184.

Eckhoff, C., Chari, S., Kromka, M., Staudner, H., Juhasz, L., Rudiger, H., and Agnish, N. 1994.

Teratogenicity and transplacental pharmacokinetics of 13-*cis*-retinoic acid in rabbits. *Toxicol. Appl. Pharmacol.* 125:34–41.

Ekker, S. C., McGrew, L. L., Lai, C.-J., Lee, J., von Kessler, D. P., Moon, R. T., and Beachy, P. 1995. Distinct expression and shared activities of members of the hedgehog gene family of *Xenopus laevis. Development* 121:2337–2347.

Ellinger-Ziegelbauer, H., and Dreyer, C. 1991. A retinoic acid receptor expressed in the early development of *Xenopus laevis. Genes Dev.* 5:94–104.

Evans, R. M. 1988. The steroid and thyroid hormone receptor superfamily. *Science* 240:889–895.

Fiorella, P. D., and Napoli, J. L. 1991. Expression of cellular retinoic acid binding protein (CRABP) in *Escherichia coli.* Characterization and evidence that holo-CRABP is a substrate in retinoic acid metabolism. *J. Biol. Chem.* 266:16572–16579.

Fujii, T., Pichel, J. G., Taira, M., Toyama, R., and Dawid, I. 1994. Expression patterns of the murine LIM class homeobox gene lim1 in the developing brain and excretory system. *Dev. Dyn.* 199:73–83.

Giguere, V. 1994. Retinoic acid receptors and cellular retinoid binding proteins: Complex interplay in retinoid signalling. *Endocr. Rev.* 15:61–79.

Giguere, V., Ong, E. S., Sequi, P., and Evans, R. 1987. Identification of a receptor for morphogen retinoic acid. *Nature* 330:624–629.

Gottlicher, M., Widmark, E., Li, Q., and Gustafsson, J. A. 1992. Fatty acids activate a chimera of clofibric acid-activated receptor and the glucocorticoid receptor. *Proc. Natl. Acad. Sci. USA.* 89:4653–4657.

Graham, A., Maden, M., and Krumlauf, R. 1991. The murine Hox-2 genes display dynamic dorsoventral patterns of expression during central nervous system development. *Development* 112:255–264.

Gudas, L. J. 1994. Retinoids and vertebrate development. *J. Biol. Chem.* 269:15399–15402.

Gudas, L. l., Sporn, M. B., and Roberts, A. B. 1994. Cellular biology and biochemistry of retinoids. In *The retinoids: Biology, chemistry, and medicine*, eds. M. B. Sporn, A. B. Roberts, and D. S. Goodman, pp. 443–520. New York: Raven Press.

Gunning, D. B., Barua, A. B., and Olson, J. A. 1993. Comparative teratogenicity and metabolism of all-*trans* retinoic acid, all-*trans* retinoyl-β-glucose and all-*trans* retinoyl-β-glucuronide in pregnant Sprague Dawley rats. *Teratology* 47:29–36.

Gustafson, A. L., Dencker, L., and Eriksson, U. 1993. Non-overlapping expression of CRBP 1 and CRABP 1 during pattern formation of limbs and craniofacial structures in the early mouse embryo. *Development* 117:451–460.

Hale, F. 1933. Pigs born without eyeballs. *J. Hered.* 24:105–106.

Happle, R., Traupe, H., Bounameaux, Y., and Fisch, T. 1984. Teratogene Wirkung von Etretinate beim Menschen. *Dtsch. Med. Wochenschr.* 109:1476–1480.

Haussler, M. R., Donaldson, C. A., Kelly, M. A., et al. 1984. Identification and quantification of intracellular retinol and retinoic acid binding proteins in cultured cells. *Biochim. Biophys. Acta* 803:54–62.

Hersh, J. H., Danhauer, D. E., Hand, M. E., and Weisskopf, B. 1985. Retinoic acid embryopathy. Timing of exposure and effects on fetal development. *J. Am. Med. Assoc.* 254:909–910.

Heyman, R. A., Mangelsdorf, D. J., Dyck, J. A., Stein, R. B., Eichele, G., Evans, R. M., and Thaller, C. 1992. 9-*cis* Retinoic acid is a high affinity ligand for the retinoid X receptor. *Cell* 68:397–406.

Ho, L., Mercola, M., and Gudas, L. 1994. *Xenopus laevis* cellular retinoic-acid binding protein: Temporal and spatial expression pattern during early embryogenesis. *Mech. Dev.* 47:53–64.

Hofmann, C., and Eichele, G. 1994. Retinoids and development. In *The retinoids: Biology, chemistry and medicine*, eds. M. B. Sporn, A. B. Roberts, and D. S. Goodman, pp. 387–441. New York: Raven Press.

Holder, N., and Hill, J. 1991. Retinoic acid modifies development of mid-brain border and affects cranial ganglion formation in zebra fish embryos. *Development* 113:1159–1170.

Howard, W. B., Willhite, C. C., Sharma, R. P., Omaye, S. T., and Hatori, A. 1989. Pharmaco-kinetics, tissue distribution and placental permeability of tetrahydro-tetramethyl-naphthalenyl-propenylbenzoic acid (a retinoidal benzoic acid derivative) in hamsters. *Eur. J. Drug Metab. Pharmacokinet.* 14:153–159.

Hunt, P., Wilkinson, D., and Krumlauf, R. 1991. Patterning the vertebrate head: Murine *Hox 2* genes mark distinct subpopulations of premigratory and migrating cranial neural crest. *Development* 112:43–50.

Husmann, M., Hoffmann, B., Stump, D. G., Chytil, F., and Pfahl, M. 1992. A retinoic acid response element from rat CRBP 1 promotor is activated by a RAR/RXR heterodimer. *Biochem. Biophys. Res. Commun.* 187:1558–1564.

Irving, D. W., Willhite, C. C., and Burk, D. T. 1986. Morphogenesis of isotretinoin-induced microcephaly and micrognathia studied by scanning electron microscopy. *Teratology* 34:141–153.

Issemann, I., and Green, S. 1990. Activation of a member of the steroid hormone receptor super-family by peroxisome proliferators. *Nature* 347:645–650.

Jarvis, B. L., Johnston, M. C., and Sulik, K. K. 1990. Congenital malformations of the external, middle and inner ear produced by isotretinoin exposure in mouse embryos. *Otolaryngol. Head Neck Surg.* 102:391–401.

Jiang, H., Gyda, M., Harnish, D. C., Chandraratna, R. A., Soprano, K. J., Kochhar, D. M., and Soprano, D. R. 1994. Teratogenesis by retinoic acid analogs positively correlates with elevation of retinoic acid receptor-beta2 mRNA levels in treated embryos. *Teratology* 50:38–43.

Jorgensen, M. B., Kristensen, H. K., and Buch, N. H. 1964. Thalidomide-induced aplasia of the inner ear. *J. Laryngol. Otol.* 78:1095–1101.

Kalin, J. R., Wells, M. J., and Hill, D. L. 1984. Effects of phenobarbital, 3-methylcholanthrene, and retinoid pretreatment on disposition of orally administered retinoids in mice. *Drug Metab. Dispos.* 12:63–67.

Kassis, I., Suderji, S., and Abdul-Karim, R. 1985. Isotretinoin (Accutane) and pregnancy. *Teratology* 32:145–146.

Kastner, P., Grondona, J., Mark, M., Gansmuller, A., LeMeur, M., Decimo, D., Vonesch, J.-L., Dolle, P., and Chambon, P. 1994. Genetic analysis of RXRα developmental function: Con-vergence of RXR and RAR signaling pathways in heart and eye morphogenesis. *Cell* 78:987–1003.

Kessel, M., and Gruss, P. 1991. Homeotic transformation of murine vertebrae and concomitant alteration of Hox codes induced by retinoic acid. *Cell* 67:89–104.

Kimelman, D., and Kirschner, M. 1987. Synergistic induction of mesoderm by FGF and TGF-β and the identification of an mRNA coding for FGF in the early *Xenopus* embryo. *Cell* 51:869–877.

Kimelman, D., Christian, J. L., and Moon, R. T. 1992. Synergistic principles of development: Overlapping patterning systems in *Xenopus* mesoderm induction. *Development* 116:1–9.

Kliewer, S. A., Umesono, K., Mangelsdorf, D., and Evans, R. 1992a. Retinoid X receptor inter-acts with nuclear receptors in retinoic acid, thyroid hormone and vitamin D$_3$ signaling. *Nature* 355:446–449.

Kliewer, S. A, Umesono, K., Noonan, J., Heyman, R., and Evans, R. 1992b. Convergence of 9-cis retinoic acid and peroxisome proliferator signalling pathways through heterodimer for-mation of their receptors. *Nature* 358:771–774.

Kliewer, S. A., Umesono, K., Heyman, R. A., Mangelsdorf, D. J., Dyck, J. A., and Evans, R. M. 1992c. Retinoid X receptor COUP-TF interactions modulate retinoic acid signalling. *Proc. Natl. Acad. Sci. USA* 89:1448–1452.

Kochhar, D. M., and Penner, J. 1987. Developmental effects of isotretinoin and 4-oxo-isotretinoin: The role of metabolism in teratogenicity. *Teratology* 36:67–75.

Kochhar, D. M., Penner, J. D., and Tellone, C. 1984. Comparative teratogenic activities of two retinoids: Effects on palate and limb development. *Teratogen. Carcinogen. Mutagen.* 4:377–387.

Kochhar, D. M., Penner, J. D., and Satre, M. 1988. Derivation of retinoic acid and metabolites from a teratogenic dose of retinol (vitamin A) in mice. *Toxicol. Appl. Pharmacol.* 96:429–441.

Kochhar, D. M., Jiang, J., Penner, J. D., and Heyman, R. A. 1993. The teratogenic activity of 9-*cis*-retinoic acid. *Teratology* 47:439.

Krust, A., Kastner, P., Petkovich, M., Zelent, A., and Chambon, P. 1989. A third human retinoic acid receptor, hRAR-alpha. *Proc. Natl. Acad. Sci. USA* 86:5210–5214.

Lammer, E. J., Chen, D. T., Hoar, R. M., Agnish, N. D., Benke, P. J., Braun, J. T., Curry, C. J., Fernhoff, P. M., Grix, A. W., Lott, I. T., Richard, J. M., and Sun, S. C. 1985. Retinoic acid embryopathy. *N. Engl. J. Med.* 313:837–841.

Lampron, C., Rochette-Egly, C., Gorry, P., Dolle, P., Mark, M., Lufkin, T., LeMeur, M., and Chambon, P. 1995. Mice deficient in cellular retinoic acid binding protein II (CRABPII) or in both CRABPI and CRABPII are essentially normal. *Development* 121:539–548.

Lawson, K. A., and Pederson, R. A. 1992. Clonal analysis of cell fate during gastrulation and early neurulation in the mouse. In *Postimplantation development in the mouse*, eds. D. J. Chadwick and J. Marsh, pp. 3–26. Chichester: John Wiley & Sons.

Lee, J., von Kessler, D. P., Parks, S., and Beachy, P. A. 1992. Secretion and localized transcription suggest a role in position signaling for products of the segmentation gene *hedgehog. Cell* 71:33–50.

Lee, M.-O., Manthey, C. L., and Sladek, N. E. 1991. Identification of mouse liver aldehyde dehydrogenases that catalyze the oxidation of retinaldehyde to retinoic acid. *Biochem. Pharmacol.* 42:1279–1285.

Lee, Q. P., Juchau, M. R., and Creech Kraft, J. 1991. Microinjection of cultured rat embryos: Studies with retinol, 13-*cis*- and all-*trans*-retinoic acid. *Teratology* 44:313–323.

Lehmann, J., Jong, L., Fanjul, A., Cameron, J. A., Lu, X. P., Haefner, P., Dawson, M., and Pfahl, M. 1992. Retinoids selective for retinoid X receptor pathways. *Science* 258:1944–1946.

Leid, M., Kastner, P., Lyons, R., Nakshatri, N., Saunders, M., Zacharewski, T., Chen, J.-Y., Staub, A., Garnier, J.-M., Mader, S., and Chambon, P. 1992. Purification, cloning and RXR identity of the HeLa cell factor with which RAR or TR heterodimerizes to bind target sequences efficiently. *Cell* 68:377–395.

Leo, M. A., Kim, C.-I., Lowe, N., and Lieber, C. S. 1989. Increased hepatic retinal dehydrogenase activity after phenobarbital and ethanol administration. *Biochem. Pharmacol.* 38:97–103.

Levin, A. A., Sturzenbecker, L. J., Kazmer, S., Bosakowski, T., Huselton, C., Allenby, G., Speck, J., Kratzeisen, C., Rosenberger, M., Lovey, A., and Grippo, J. 1992. 9-*Cis* retinoic acid stereoisomer binds and activates the nuclear receptor RXRα. *Nature* 355:359–361.

Levin, M. S., Li, E., Ong, D. E., and Gordon, J. I. 1987. Comparison of tissue-specific expression and developmental regulation of two closely linked rodent genes encoding cytosolic retinol-binding protein. *J. Biol. Chem.* 262:7118–7124.

Lohnes, D., Kastner, P., Dierich, A., Mark, M., LeMeur, M., and Chambon, P. 1992. Function of retinoic acid receptor gamma in the mouse. *Cell* 73:643–658.

Lufkin, I., Lohnes, D., Mark, M., Dierich, A., Gorry, P., Gaub, M.P., Lemeur, M., and Chambon, P. 1993. High postnatal lethality and testis degeneration in retinoic acid receptor alpha mutant mice. *Proc. Natl. Acad. Sci. USA* 90:7225–7229.

MacGregor, T. M., Copeland, N. G., Jenkins, N. A., and Giguere, V. 1992. The murine gene for cellular retinoic acid binding protein type II. Genomic organization, chromosomal localization, and post-transcriptional regulation by retinoic acid. *J. Biol. Chem.* 276:7777–7783.

Maden, M. 1994a. Vitamin A in embryonic development. *Nutr. Rev.* 52:s3–s12.

Maden, M. 1994b. Distribution of cellular retinoic acid-binding proteins I and II in the chick embryo and their relationship to teratogenesis. *Teratology* 50:294–301.

Maden, M., Ong, D. E., Summerbell, D., and Chytil, F. 1988. Spatial distribution of cellular protein-binding to retinoic acid in the chick limb bud. *Nature* 335:733–735.

Magendie, F. 1816. Sur les proprietes nutritives des substances qui ne contiennent pas d'azote *Ann. Chim. Phys.* (1)3:66–77.

Mahmood, R., Morriss-Kay, G. M., and Flanders, K. C. 1993. Transforming growth factor βs in early mouse embryogenesis and their relationship with retinoids. In *Retinoids in research and clinical applications*, eds. M. A. Livrea and L. Packer, pp. 397–407. New York: Marcel Dekker.

Mangelsdorf, D. 1994. Vitamin A receptors. *Nutr. Rev.* 52:s32–s44.

Mangelsdorf, D. J., Ong, E. S., Dyck, J. A., and Evans, R. M. 1990. Nuclear receptor that identifies a novel retinoic acid response pathway. *Nature* 345:224–229.

Marklew, S., Smith, D., Mason, C. S., and Old, R. W. 1994. Isolation of a novel RXR from *Xenopus* that most closely resembles mammalian RXRβ and is expressed throughout early development. *Biochim. Biophys. Acta.* 1218:267–272.

Marshall, H., Nonchev, S., Sham, M. H., Muchamore, I., Lumsden, A., and Krumlauf, R. 1992. Retinoic acid alters hindbrain Hox code and induces transformation of rhombomeres 2/3 into 4/5 identity. *Nature* 360:737–741.

McCollum, E. V., and Davis, M. 1913. The necessity of certain lipids in the diet during growth. *J. Biol Chem.* 15:167–175.

Mohler, J., and Vani, K. 1992. Molecular organization and embryonic expression of the *hedgehog* gene involved in cell-cell communication in segmental patterning of *Drosophila*. *Development* 115:957–971.

Morriss, G. M. 1972. Morphogenesis of the malformations induced in rat embryos by maternal hypervitaminosis A. *J. Anat.* 113:241–250.

Morriss, G. M., and Thorogood, P. V. 1978. An approach to cranial neural crest cell migration and differentiation in mammalian embryos. In *Development in mammals*, ed. M. F. Johnson, vol. 3, pp. 363–441. Amsterdam: Elsevier-North Holland.

Morriss-Kay, G. M. 1993. Retinoic acid and craniofacial development: Molecules and morphogenesis. *Bioessays* 15:9–15.

Morriss-Kay, G. M., Murphy, P., Hill, R., and Davidson, D. 1991. Effects of retinoic acid excess on expression of Hox 2.9 and Krox-20 and on morphological segmentation of the hindbrain of mouse embryos. *EMBO J.* 10:2985–2995.

Muerhoff, A. S., Griffen, K. J., and Johnson, E. F. 1992. The peroxisome proliferator-activated receptor mediates the induction of CYP A46, a cytochrome P450 fatty acid ω-hydroxylase, by clofibric acid. *J. Biol. Chem.* 267:19051–19053.

Napoli, J. L. 1993. Biosynthesis and metabolism of retinoic acid: Roles of CRBP and CRABP in retinoic acids homeostasis. *J. Nutr.* 123:362–366.

Napoli, J. L., Posch, K. P., Fiorella, P. D., and Boermann, M. H. E. M. 1991. Physiological occurrence, biosynthesis and metabolism of retinoic acid: Evidence for roles of cellular retinol-binding protein (CRBP) and cellular retinoic acid-binding protein (CRABP) in the pathway of retinoic acid homeostasis. *Biomed. Pharmacother.* 45:131–143.

Napoli, J. L., Posch, K. P., Fiorella, P. D., Boerman, M. H. E. M., Saerno, G. J., and Burns, R. D. 1993. Roles of CRBP and CRABP in the metabolic chanelling of retinoids. In *Retinoids: Progress in research and clinical applications*, eds. M. A. Livrea and L. Packer, pp. 29–48. New York: Marcel Dekker.

Niazi, I. A., and Saxena, S. 1968. Inhibitory and modifying influence of excess of vitamin A on tail regeneration in *Bufo* tadpoles. *Experientia* 24:852–853.

Ong, D. E., Newcomer, M. E., and Chytil, F. 1994. Cellular retinoid-binding proteins. In *The retinoids: Biology, chemistry, and medicine*, eds. A. B. Sporn and D. S. Goodman, pp. 283–317. New York: Raven Press.

Osborne, T. B., and Mendel, L. B. 1913. The influence of butter-fat on growth. *J. Biol. Chem.* 16:423–437.

Osmundsen, H., Bremer, J., and Pedersen, J. I. 1991. Metabolic aspects of peroxisomal-β-oxidation. *Biochem. Biophys. Acta* 1085:141–158.

Papalopulu, N., Clarke, J. D., Bradley, L., Wilkinson, D., Krumlauf, R., and Holder, N. 1991. Retinoic acid causes abnormal development and segmental patterning of the anterior hindbrain in *Xenopus* embryos. *Development* 113:1145–1158.

Petkovich, M., Brand, N. J., Krust, A., and Chambon, P. 1987. A human retinoic acid receptor which belongs to the family of nuclear receptors. *Nature* 330:444–450.

Petkovich, M. 1992. Regulation of gene expression by vitamin A: The role of the nuclear retinoic acid receptors. *Annu. Rev. Nutr.* 12:443–471.

Pignatello, M. A., Kauffman, F., and Levin, A. A. 1995. Multiple factors contribute to the toxicity of aromatic retinoid TTNPB (Ro 13-7410). *Toxicologist* 15(1):159.

Pijnappel, W. W. M., Hendriks, H. F. J., Folkers, G. E., van den Brink, C. E., Dekker, E. J., Edellenbosch, C., van der Saag, P. T., and Durston, A. J. 1993. The retinoid ligand 4-oxo-retinoic acid is a highly active modulator of positional specification. *Nature* 366:340–344.

Poellinger, L., Goettlicher, M., and Gustafsson, J.-K. 1992. The dioxin and peroxisome proliferator-activated receptors: Nuclear receptors in search of endogenous ligands. *TIPS* 13:241–244.

Riddle, R., Johnson, R., Lauter, E., and Tabin, C. 1993. *Sonic hedgehog* mediates the polarizing activity of ZPA. *Cell* 75:1401–1416.

Rosa, F. W. 1983. Teratogenicity of isotretinoin. *Lancet* 2:513.

Rosa, F. 1990. FDA joint fertility and maternal health and dermatologic drug advisory committee. May 21. United States Food and Drug Administration, 5600 Fishers Lane (HFD 733), Rockville, MD.

Ross, A. C. 1993. Cellular metabolism and activation of retinoids. Roles of cellular retinoid-binding proteins. *FASEB J.* 7:317–327.

Rossant, J., Zirngibl, R., Cado, D., Shago, M., and Giguere, V. 1991. Expression of a retinoic acid response element-*hsplacZ* transgene specific domains of transcriptional activity during mouse embryogenesis. *Genes Dev.* 5:1333–1344.

Ruberte, E., Dolle, P., Chambon, P., and Morriss-Kay, G. M. 1991. Retinoic acid receptors and cellular retinoid binding proteins. II. Their differential pattern of distribution during morphogenesis in mouse embryos. *Development* 111:46–50.

Ruberte, E., Friederich, V., Morriss-Kay, G. M., and Chambon, P. 1992. Differential distribution patterns of CRABP I and CRABP II transcripts during mouse embryogenesis. *Development* 115:973–989.

Ruiz i Altaba, A., and Jessell, T. M. 1991. Retinoic acid modifies the pattern of cell differentiation in the central nervous system of neurula stage *Xenopus* embryos. *Development* 112:945–958.

Sandberg, J. A., Eckhoff, C., Nau, H., and Slikker, W. 1994. Pharmacokinetics of 13-*cis*-, all-*trans*-, 13-*cis*-4-oxo-, and all-*trans*-4-oxo-retinoic acid after intravenous administration in the cynomolgus monkey. *Drug Metab. Dispos.* 22:154–160.

Sass, J. O. 1994. 3,4-Didehydroretinol may be present in human embryos/fetuses. *Reprod. Toxicol.* 8:191.

Satre, M. A., Penner, J. D., and Kochhar, D. M. 1989. Pharmacokinetic assessment of teratologically effective concentrations of an endogenous retinoic acid metabolite. *Teratology* 39:341–348.

Schardein, J. L. 1993. Human studies: retinoic acid embryopathy. In *Chemically induced birth defects*, ed. J. L. Schardein, pp. 558–567. New York: Marcel Dekker.

Schmidt, A., Endo, N., Rutledge, S. J., Vogel, R., Shinar, D., and Rodan, G. A. 1992. Identification of a new member of the steroid hormone superfamily that is activated by a peroxisome proliferator and fatty acids. *Mol. Endocrinol.* 6:1634–1641.

Schuele, R., Ranagarajan, P., Yanf, N., Kliewer, S., Ransone, L. J., Bolado, J., Verma, I. M., and Evans, R. M. 1991. Retinoic acid is a negative regulator of AP-1 responsive genes. *Proc. Natl. Acad. Sci. USA* 88:6092–6096.

Scott, W., J., Walter, R., Tzimas, G., Sass, J. O., Nau, H., and Collins, M. 1994. Endogenous status of retinoids and their cytosolic binding proteins in limb buds of chick versus mouse embryos. *Dev. Biol.* 165:397–409.

Sharpe, C. R. 1992. Two isoforms of retinoic acid receptor α expressed during *Xenopus* development respond to retinoic acid. *Development* 39:81–93.

Shawlot, W., and Behringer, R. R. 1995. Requirement for *Lim1* in head-organizer function. *Nature* 374:425–430.

Simone, A., Acampora, D., Arcioni, L., Andrews, P. W., Boncinelli, E., and Mavilio, F. 1990. Sequential activation of *HOX2* homeobox genes by retinoic acid in human embryonal carcinoma cells. *Nature* 346:763–766.

Sive, H. L., and Chen, P. F. 1991. Retinoic acid perturbs the expression of Xhox.lab genes and alters mesodermal determination in *Xenopus laevis. Genes Dev.* 5:1321–1332.

Sive, H., Draper, B., Harland, R.M., and Weintraub, H. 1990. Identification of a retinoic acid sensitive period during axis formation in *Xenopus laevis. Genes Dev.* 4:932–942.

Slack, J. M. W., and Tannahill, D. 1992. Mechanism of anteroposterior axis specification in vertebrates. Lessons from amphibians. *Development* 114:285–302.

Smith, W. C., Nakshatri, H., Leroy, P., and Chambon, P. 1991. A retinoic acid response element is present in the mouse cellular retinol binding protein 1 (mCRBP1) promoter. *EMBO J.* 10:2223–2230.

Soprano, D. R., Gyda, M., Jiang, H., Harnish, D., Ugen, K., Satre, M. Chen, L., Soprano, K., and Kochhar, D. M. 1994. A sustained elevation in retinoic acid receptor-β2 mRNA and protein occurs during retinoic acid-induced fetal dysmorphogenesis. *Mech. Dev.* 45:243–253.

Spemann, H., and Mangold, H. 1924. Ueber Induktion von Embryonalanlagen durch Implantation artfremder Organisatoren. *Wilhelm Roux Arch. Entw. Mech. Org.* 100:599–638.

Sporn, M. D., Roberts, A. B., and Goodman, D. S., eds. 1994. *The retinoids: Biology, chemistry and medicine*, 2nd ed. New York: Raven Press.

Stornaivolo, A., Acapora, D., Pannese, M., D'Esposito, M., Morelli, F., Migliaccio, E., Rambaldi, M., Failla, A., Nigro, V., Simone, A., and Boncinelli, E. 1990. Human Hox genes are differentially activated in embryonal carcinoma cells according to their position within four loci. *Cell Differ. Dev.* 31:119–127.

Sucov, H. M., Murakami, K. K., and Evans, R. M. 1990. Characterization of an autoregulated response element in the mouse retinoic acid receptor type β gene. *Proc. Natl. Acad. Sci. USA* 87:5392–5396.

Sucov, H. M., Dyson, E., Gumeringer, C. L., Price, J., Chien, K. R., and Evans, R. 1994. RXRα mutant mice establish a genetic basis for vitamin A signaling in heart morphogenesis. *Genes Dev.* 8:1007–1018.

Sulik, K. K., Cook, C. S., and Webster, W. S. 1988. Teratogens and cranial malformations: Relationships to cell death. *Development* 103:213–232.

Summerbell, D., and Waterson, N. 1993. Retinoic acid, an autocatalytic morphogen. In *Retinoids. Progress in research and clinical applications*, eds. M. A. Livrea and L. Packer, pp. 439–451. New York: Marcel Dekker.

Tabata, T., Eaton, S., and Kornberg, T. B. 1992. The *Drosophila hedgehog* gene is expressed specifically in posterior compartment cells and is a target of engrailed regulation. *Genes Dev.* 6:2635–2645.

Tabin, C. J. 1991. Retinoids, homeoboxes and growth factors: Towards molecular models for limb development. *Cell* 66:199–217.

Taira, M., Jamrich, M., Good, P. J., and Dawid, I. B. 1992. The LIM domain-containing homeobox gene *Xlim-1* is expressed specifically in the organizer-region of *Xenopus* gastrula embryos. *Genes Dev.* 6:356–366.

Taira, M., Otani, H., Saint-Jeannet, J., and Dawid, B. 1994a. Role of the LIM class homeodomain protein *Xlim-1* in neural and muscle induction by the Spemann organizer in *Xenopus. Nature* 372:677–679.

Taira, M., Otani, H., Jamrich, M., and Dawid, I. 1994b. Expression of the LIM class homeobox gene *Xlim-1* in pronephros and CNS cell lineages of *Xenopus* embryos is affected by retinoic acid and exogastrulation. *Development* 130:1525–1538.

Thaller, C., and Eichele, G. 1987. Identification and spatial distribution of retinoids in the developing chick limb bud. *Nature* 327:625–628.

Thaller, C., and Eichele, C. 1990. Isolation of 3,4-didehydroretinoic acid, a novel morphogen in the chick limb bud. *Nature* 345:815–819.

Tickle, C., and Brickell, P. 1991. Retinoic acid and limb development. *Semin. Dev. Biol.* 2:189–197.

Tzimas, G., Sass, J. O., Wittfoht, W., Elmazar, M. A., Ehlers, K., and Nau, H. 1994. Identification of 9,13-di-*cis*-retinoic acid as a major plasma metabolite of 9-*cis*-retinoic acid and limited transfer of 9-*cis*-retinoic acid and 9,13-di-*cis*-retinoic acid to the mouse and rat embryos. *Drug Metab. Dispos.* 22:928–936.

Underwood, B. A. 1993. The epidemiology of vitamin A deficiency and depletion (hypovitaminosis A) as a public health problem. In *Retinoids. Progress in research and clinical applications*, eds. M. A. Livrea and L. Packer, pp. 171–184. New York: Marcel Dekker.

Willhite, C. C. 1993. Cellular retinoic acid-binding proteins in the pathogenesis of retinoid congenital malformations. In *Retinoids. Progress in research and clinical applications*, eds. M. A. Livrea and L. Packer, pp. 409–437. New York: Marcel Dekker.

Willhite, C. C., and Dawson, M. I. 1990. Structure-activity relationships of retinoids in developmental toxicology IV. Planar cisoid conformational restriction. *Toxicol. Appl. Pharmacol.* 103:324–244.

Willhite, C. C., Hill, R. M., and Irving, D. W. 1986. Isotretinoin-induced craniofacial malformations in humans and hamsters. *J. Craniofac. Genet. Dev. Biol.* 2(suppl.):193–209.

Willhite, C. C., Wier, P. J., and Berry, D. 1989. Dose response and structure activity considerations in retinoid-induced dysmorphogenesis. *Crit. Rev. Toxicol.* 20:113–135.

Willhite, C. C., Jurek, A., Sharma, R., and Dawson, M. 1992. Structure-activity relationships of retinoids with embryonic cellular retinoic acid-binding protein. *Toxicol. Appl. Pharmacol.* 112:144–153.

Willhite, C. C., Dawson, M., and Reichert, U. 1996. Receptor-selective retinoid agonists and teratogenic activity. *Drug Metab. Rev.* 28(1–2):105–119.

Wolbach, S. B., and Howe, P. R. 1925. Tissue changes following deprivation of fat soluble A vitamin. *J. Exp. Med.* 42:753–777.

Wolf, G. 1984. Multiple functions of vitamin A. *Physiol. Rev.* 64:873–937.

Wolpert, L. 1969. Positional information and spatial pattern of cellular differentiation. *J. Ther. Biol.* 25:1–47.

Yang-Yen, H. F., Chambard, J. C., Sun, Y. L., Schmidt, T. J., Drouin, J., and Karin, M. 1990. Transcriptional interference between *cJun* and glucocorticoid receptor: Mutual inhibition of DNA binding due to protein–protein interaction. *Cell* 62:1205–1215.

Yu, V. C., Delsert, C., Anderson, B., Holloway, J., Devary, O., Naeaer, Kim, S.Y., Boulin, J.-M., Glass, C. K., and Rosenfeld, M. G. 1991. RXRβ: A coregulator that enhances binding of retinoic acid, thyroid hormone, and vitamin D receptors to their cognate response elements. *Cell* 67:1251–1266.

Zhang, X.-K., Lehmnan, J., Hoffmann, B., Dawson, M., Cameron, J., Graupner, G., Hermann, T., Tran, P., and Pfahl, M. 1992a. Homodimer formation of retinoid X receptor induced by 9-*cis*-retinoic acid. *Nature* 358:589–591.

Zhang, X.-K., Hoffmann, B., Tran, P., Graupner, G., and Pfahl, M. 1992b. Retinoid X receptor is an auxiliary protein for thyroid hormone and retinoic acid receptors. *Nature* 355:441–445.

Zhang, B., Marcus, S. L., Sajjadi, F. G., Alvars, K., Reddy, J. K., Subramani, R. A., and Capone, J. P. 1992c. Identification of a peroxisome proliferator-responsive element upstream of a gene encoding rat peroxisomal enoyl-CoA hydratase/3-hydroxyacyl-CoA dehydrogenase. *Proc. Natl. Acad. Sci. USA* 89:7541–7545.

Chapter 3

Toxicologic Problems in Pediatric Populations in Eastern Europe

Charles V. Smith

Infants and children are confronted with many complex problems throughout the world. However, the long-term deterioration of political, economic, and social conditions, followed by the recent rather abrupt changes occurring in the countries of Eastern Europe, have produced some problems that are common to all economically limited civilizations, but some problems that are unique to the children of these countries. The topic of this chapter is not one on which I view myself as an expert, and the reader is cautioned to accept the present chapter as an effort to increase the recognition by western scientists of some of the relevant difficulties facing the children of Eastern Europe. For the purposes of the present discussion, we include the adjacent newly independent states of the former Soviet Union, but events in Albania are not included because of a general lack of information regarding that country's situation.

The decisions that need to be reached include the requirements of the region as a whole, and include the needs of the countries and villages to balance their immediate needs for economic stability and transformation to free-market structures with the importance of building stable foundations for long-term growth and quality of life. The efforts to manage the compromises between present and future needs are perhaps most acute in issues of environmental protection, where there often exists a perceived need to compromise environmental protection in an effort to maximize production of food, energy, and economic growth.

Many of the difficulties faced by the Eastern European countries today are consequences of the choices made in years past. Particularly striking are the enormous costs, both economic and human, that have been imposed by environmental compromises of past generations. These compromises resemble some of the issues under discussion in the United States and other relatively affluent

51

countries at the present time. Consideration of the history and the present situations confronting the people of Eastern Europe may help us to better recognize the potential outcomes and consequences of our own choices.

ENVIRONMENT

The potential hazards associated with exposures to trace amounts of environmental pollutants are of less concern to someone who has to dodge sniper fire to get to a hospital, or has to dash, under fire, from that hospital just before it is destroyed by artillery (Cermerlic & Schaller, 1995). One published report from Croatia (Metelko et al., 1992) documented a remarkable flexibility in the capacity of the local health care system to sustain insulin delivery and disease monitoring during the first year or so of the war in that region. The continued conflict has no doubt stretched that system beyond its capabilities, but the adaptability of the health care system is a function of the people and their levels of commitment and that should endure. The acute and longer lasting problems associated with being a child in the middle of an armed conflict exceed the scope of the present review, but these problems must be kept in mind. The children who do survive these wars will need to catch up on missed educational opportunities and social development and will have to work in cooperation with the children of the people who killed their parents and denied them their childhood, if the region is to improve environmental health and quality of life in the region (Table 3-1). We can only hope that these children prove to be wiser than their parents have thus far.

Although a healthy environment is an essential component of a healthy future for any group, environmental pollution is perhaps more significant in pediatric populations (Table 3-2) because they have a longer life expectancy before them (von Muhlendahl et al., 1995). In addition, infants and children may be more sensitive to some toxic substances than are adults. Finally, the consequences of many past and present environmentally damaging activities can only be reversed with considerable difficulty, if at all. Although some may find a certain level of satisfaction in forcing their children to have to clean up after

Table 3-1   Objectives of the United Nations Conference on Environment and Development

1 Accelerated evaluation of risks from chemicals to human health and the environment.
2 International coordination of classification and labeling of chemicals.
3 Improved exchange of necessary information on risks and responses.
4 Development of safer chemicals and methods of use.
5 Assist developing and transitional countries in evaluation of environmental and health hazards. Assist in enforcement of legislation, as appropriate.
6 More effective regulation of international traffic in hazardous chemicals.

Note: Adapted from "Health in Central Europe," by F. Gutzweiller, 1991, *Ann. Nutr. Metab.* 35(suppl. 1):64–68, copyright © 1991 Karger, Basel.

Table 3-2 Environmental Problems of Special Relevance to European Children and Youth

Global problems	European problems
Overpopulation	Alcohol
Poverty	Leaded gasoline
Hunger	Drugs
Epidemic disease	Wood preservatives
Radioactive wastes	Indoor air pollution
Chemical wastes	Noise
	Psychosocial stresses
	Sulfur dioxide
	Tobacco smoke
	Traffic

Note: Adapted from von Muhlendahl and Otto (1994), with permission.

them, dirty socks are hardly a fair balance with radioactive waste, heavy metal contamination, and pesticides in the drinking water.

One of the most heavily polluted regions in Eastern Europe is the Upper Silesian region around Katowice in southern Poland. The area is plagued by the release of untreated industrial waste and human sewage into surface waters and dust and other airborne emissions from industry and home use. Additional health stresses arise from smoking, alcohol consumption, and abuse of other drugs, as well as the malnutrition and other factors arising from poverty. Other specific environmental contaminants identified thus far include benzo[a]pyrene and other polycyclic aromatic hydrocarbons, phenol, formaldehyde, SO_2, N_2O_5, Pb, Cd, and Cu, as well as dust and related pollutants (Dutkiewicz, 1991; Balter, 1993; Norska-Borowka, 1994). Reports suggest a frequency of premature or small for gestational age birth of up to 20%, and infant mortality is reported to be as high as 53 per 1000 live births in Rozbark (Norska-Borowka, 1994).

Studies of heavy metal content in fish and game meats indicate contamination by cadmium from environmental sources (Falandysz, 1994; Krelowska-Kulas, 1995). Other studies indicate more extensive contamination of food supplies by heavy metals including lead and mercury, with nutritional deficiencies of other metals such as iron and selenium (Dobrowolski & Smyk, 1993). In addition, microbiological contamination of the food is a significant problem and improvements in monitoring and protection of the food chain are needed. Poland continues to be stressed by an epidemic of food poisoning associated with *Salmonella enteritidis*, which began over a decade ago (Glosnicka & Kunikowska, 1994).

One method of minimizing the problems of environmental contamination is through recycling of materials such as industrial byproducts, thus converting wastes into useful materials. Although this may seem like a win/win plan, a problem arises when waste traders shift toxic materials from developed countries to developing or transitional countries in which regulations are less restric-

Table 3-3   Issues in Environmental Protection

1 Personal responsibility
2 Economically viable options
3 Governmental oversight
  a Emissions monitoring
  b Technical expertise
    (1) Assist rational prioritization
    (2) Interpret data from monitoring and continuing basic research
4 International and interagency cooperation

tive or are more readily influenced to meet short-term economic goals, rather than providing environmentally sound procedures that can be sustained in the long run. In 1990, some 40 million tons of waste were exported for "recycling," generating a revenue of approximately $38 billion (Mackenzie, 1993). In the guise of "recycling," many toxic pollutants may simply be dumped. It is reasonable to expect that there will be a tendency for toxic wastes to be moved from richer countries to poorer ones. Although personal responsibility alone would do much for preventing abuses in the system, economically feasible alternatives for earning a living would permit optimization of environmental decisions (Table 3-3). However, regulation and enforcement are necessary to prevent abuses of the poor and relatively defenseless.

A recent study of age-dependent changes in cloning efficiencies and mutation frequencies in the *hprt* genes of T lymphocytes in a Russian population of unknown smoking status revealed greater mutation frequencies in the Russian cohort than were observed in a population of smokers from the United States (Jones et al., 1995). This difference could not be attributed to a greater proportion of smokers in the Russian group, although greater average intensities of smoking or greater susceptibilities of the individuals could contribute to the observed differences. However, the most reasonable interpretation of the data would seem to be that they reflect significantly greater environmental exposures to genotoxic agents.

HEALTH CARE

One approach to the consideration of health care issues begins with the prenatal and perinatal care surrounding the gestation and birth of the children. In this respect, the transitional countries of Eastern Europe show rates of infant mortality that exceed those of Western European countries by about 100% (Nanda et al., 1993), although considerable variation is seen in countries within the region. As is illustrated in Figure 3-1, there is a tendency for infant mortality rates to be lower in countries with longer life expectancies, but some countries, particularly the newly independent states (NIS) of the former Soviet Union, show infant mortality rates that are disproportionately high. Perhaps more discouraging is the fact that this gap has widened over the last decade. In addition, maternal

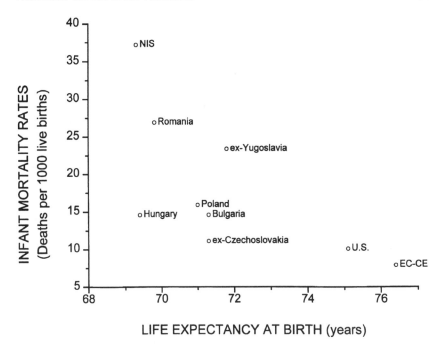

Figure 3-1 Infant mortality and life expectancy at birth in industrialized countries. The data are for 1990 and were adapted from Nanda et al. (1993).

mortalities in Eastern Europe have increased from approximately three times in 1980 to almost six times the rates in Western Europe by 1990. Even within countries, marked differences exist. The infant mortality rate in Hungary in 1970 was 36 out of 1000 live births, but in Hungarian gypsies the rate was 116 per 1000. By 1990, the national rate had dropped to 14.8, and infant mortality among the gypsies was 22 out of 1000 live births (Schuler, 1992).

Until the late 1980s, national health care in Europe followed three basic schemes: the Breveridge model, exemplified by the United Kingdom, the Bismarck sickness insurance model, and the Shemashko model of state-directed health care established in the former Soviet Union and followed by most of the Soviet bloc countries after the Second World War (Vienonen & Wlodarczyk, 1993). Since the political reversals of the last decade, many of the Eastern European countries and newly independent states of the former Soviet Union have declared their intents to discard the Shemashko system and adopt national health care programs more similar to the Bismarck sickness insurance design. The German system offers greater control of health care delivery through the continuing negotiation of values and services that result in a greater degree of control over the system by physicians and health care institutions. On the other hand, this system results in higher costs and a greater tendency for overtreatment and for treatment rather than prevention. The need for cost control and effectiveness will require the countries of Central and Eastern Europe/newly independent states to emphasize prevention and carefully selected health care priorities, because uncontrolled medical practices would exhaust a health care budget in a very short time. In the Russian Federation, the state collects 8–12% of workers' salaries as premiums, and the state covers children, the elderly, and the disabled. In addition to numerous deficiencies, the relatively low pay of physicians in the region does not provide a sound base for establishing an incentive system of better pay for more work in the delivery of health care (Vienonen & Wlodarczyk, 1993). Many problems remain, but at least the process of health care reform in the region has begun.

Another feature of the attempts of these regions to adopt health care reform is that health care reforms have only been launched successfully by strong governments, whereas the new governments of Eastern Europe generally are anything but strong. Many segments of the societies of the nations of the region are insecure about the manner in which they will be treated during the process of transformation of health care. Without support of the population for reform, leaders offering simplistic solutions to complex problems will find their audiences, perhaps at the expense of efforts to build the solid economic and social foundations needed for real growth in the area.

Among the needs for improved health care in Eastern Europe are coordination of information, assistance in responding to local crises and shortages, particularly in vaccines, drugs, and medical equipment, assistance in the development of national health policies, and aid in efforts to encourage healthier lifestyles and other preventative measures (Table 3-1) (Danzon & Litinov, 1993). In addition to governmental efforts, independent attempts to coordinate information and exchange regarding environmental problems of pediatric populations have been initiated (von Muhlendahl & Otto, 1994).

Infectious diseases are the major causes of death worldwide, and although many of the diseases are primarily associated with tropical distributions, environmental contamination, poor sanitation, and inadequate health care set the stage for chronic or epidemic increases in the incidences of these diseases. Ten years ago nearly 5 million children died each year of diseases readily preventable by immunization (Grant & Nakajima, 1993). Over the period of 1980–1990, immunization coverage of children in developing countries increased from approximately 20% to 80%. The World Health Organization estimates that 3.2 million young lives were saved from measles, neonatal tetanus, and pertussis in 1990 alone. Despite the remarkable progress, infectious diseases remain a major cause of death worldwide (Table 3-4) and another 2.1 million children still die each year from diseases from the target list (Table 3-5).

An additional 7–8 million deaths occur each year from bacterial and viral diseases that could be prevented through improvements in existing vaccines and development of vaccines that are within the limits of present technologies. These advances also could prevent an estimated 900 million severe illnesses

Table 3-4 Death from Infectious Diseases, Worldwide

Acute respiratory infections	6,900,000
Diarrheal diseases	4,200,000
Tuberculosis	3,300,000
Malaria	1,000,000–2,000,000
Hepatitis	1,000,000–2,000,000

Note: Data for the year 1990 (Gibbons, 1992).

among children annually (Hartvelt, 1993). The development of effective vaccines against malaria and other important diseases could prevent millions of additional deaths.

Although vaccination generally is high in Eastern Europe, pockets of nonimmunized people that build up over a period of 5–6 years can lead to outbreaks of disease (Rooure & Oblapenko, 1993). The incidence rates of many infectious diseases in the area reflect such regional outbreaks, and a coordinated ability of the governments to monitor and identify these outbreaks and respond appropriately is a much-needed complement to primary prophylactic vaccination programs. Although the number of cases of pulmonary tuberculosis has remained relatively stable over the last few years, the 117,569 cases reported by countries in the region in 1992 do not reflect a minor problem. Nevertheless, 13 countries in the region have stopped immunization with the BCG vaccine (Rooure & Oblapenko, 1993). The patterns of these communicable diseases in Eastern Europe generally reflect similar patterns of hot spots, occasionally flaring into outbreaks of greater intensity. Polio hot spots exist in the Azerbaijan, whereas a recent intensive immunization program in Romania has reduced poliomyelitis from 25 cases in 1991 to only 2 cases in 1992 (Rooure & Oblapenko, 1993). Azerbaijan, the Ukraine, and adjacent countries also have seen a recent epidemic in diphtheria, apparently due to low and decreasing immunization coverage rates among infants and children, the lack of immunity in adults, in part due to lower exposure-derived opportunities to reinforce existing immunity, and the movements of large populations during the last few years (Rooure & Oblapenko, 1993; Anonymous, 1995). Hepatitis B surface antigen positivity rates are high in the area, with as many as 8% of the people in some regions responding.

Recent studies in cell biology and immunology suggest that chronic activation of acute-phase responses results in poor linear growth and failure to attain

Table 3-5 Diseases Targeted by WHO for Immunization Coverage

Measles	Tetanus
Diphtheria	Poliomyelitis
Pertussis	Tuberculosis (BCG)

Note: Adapted from "The Children's Vaccine Initiative," by F. Hartvelt, 1993, *World Health* 46:4–6, with permission.

full genetic potential. Poor growth is associated with impaired cognitive development and behavioral abnormalities in young children, with resultant compromised productivity of affected individuals in adulthood (Solomons et al., 1993). Most of the phenotypic changes caused by chronic immunostimulation are mediated by leukocyte cytokines, which function directly or indirectly to divert dietary nutrients away from tissue growth to metabolic processes that support the immune response and defense functions. In population studies, laboratory analyses have revealed infections in large portions of the populations studied who did not have otherwise evident indications of infection. The "malnutrition–infection complex," which is a major problem in developing countries, is likely to be a problem in Eastern Europe that has not been appreciated fully as yet, and one whose impact on the area has the potential to increase if health and sanitation conditions deteriorate during economic crises or degeneration. The longer term effects of compromised intellectual development of affected children, coupled with an increase in the number of behaviorally maladjusted individuals, would initiate a tragic downward spiral from which it would be increasingly difficult for a society to extricate itself. Other environmental factors also appear to be capable of influencing growth responses in similar ways, but the processes involved have not been studied as extensively as yet. The principles of neurodevelopmental outcomes of chronic toxicities will need extensive additional study to guide these continuing efforts.

In a recent study of the health status of children adopted from Romania (Johnson et al., 1992), only 15% of 65 adoptees brought to the United States were judged to be physically healthy and developmentally normal. This finding is even more remarkable in view of the fact that these children had been selected by their adoptive parents or their representatives, and presumably by other persons and agencies, as the infants and children who were the most healthy, at least in appearance. Of the children examined, over half showed evidence of past or present hepatitis B infection and one-third of the total were infected by intestinal parasites. None of these children were found to be HIV positive, although inaccuracies in the reports of HIV screening in Romanian children who were candidates for foreign adoption have been reported (Kurtz, 1991; Jenista, 1992). The problem of pediatric HIV infection in Romania appears to be unique to that country and resulted from horizontal spread of the virus, apparently from the use of unsterilized needles and contaminated blood products in efforts by health care personnel to compensate for the effects of extreme malnourishment in children in Romanian orphanages (Patrascu & Dumi-trescu, 1993).

Another special problem facing pediatric populations in Eastern Europe is the continuing influence of the nuclear power plant disaster at Chernobyl. Even though the accident occurred more than 10 years ago (April 26, 1986), the continuing contamination of the area and the continuing demands of dealing with the recognized and potential physical health problems arising from human exposure continue to burden the limited resources of the region. Increased rates of thyroid cancer in children in the Ukraine have been suggested in recent reports, from 0.04 to 0.06 per 100,000 children during 1981 to 1985 to rates of 0.19 to 0.35 per 100,000 in 1990–1992 (Kuchuk, 1994). One report indicates that the incidence of thyroid cancer in children under the age of 15 years increased from about 0.2 per 100,000 children in the mid 1980s to more than 10 per 100,000 in 1991 (Baverstock, 1993). Some of the apparent increase in rates of thyroid cancer might reflect increased screening intensity, because autopsy examinations frequently reveal hidden thyroid tumors that would not have been found without specific examination of serial histological sections. Nevertheless, studies of the association of radiation exposure with thyroid cancer have revealed:

1 Infants and young children are up to 10 times more sensitive to radiation-induced thyroid cancer than are adults.
2 There does not appear to be a dose threshold for radiation exposure, and risk for disease increases linearly with increased dose of radiation.
3 The lag time between radiation exposure and observations of first cancers can be as short as 3–4 years.

In general, other studies seeking correlations between disease increases in the region and radiation exposure resulting from the Chernobyl disaster have not uncovered strong evidence of the anticipated associations (Weinberg et al., 1995). The long latency period commonly observed between exposure and expression of most types of cancer would suggest that the true effects of Chernobyl will not be seen for several more years. In addition, the overall health status of the people living in the Ukraine and surrounding regions has been declining in recent years, so that efforts to monitor changes in health outcomes will be complicated by what will probably continue to be a shifting baseline. In the meantime, the psychological burden of living with the expectation of inevitable disease caused by radiation exposure is causing clinically evident stress in many of the people exposed as a result of the accident (Weinberg et al., 1995).

LIFESTYLE

As a general rule, life expectancy increases with economic development (Smith & Marmot, 1991). A comparison of the life expectancies at birth for men and women in countries of Eastern Europe with the life expectancies in Western Europe, Japan, and the United States is presented in Figure 3-2. In addition to the trend for life expectancies of males and females in a given country to correlate, and the 5–9 year greater life expectancy of females in each country, the data in Figure 3-2 show a separation into two distinct populations. With increasing economic development, the major causes of death shift from infections, malnutrition, and childhood diseases toward chronic degenerative diseases such as cancer and cardiovascular disease. The economic stagnation and regression

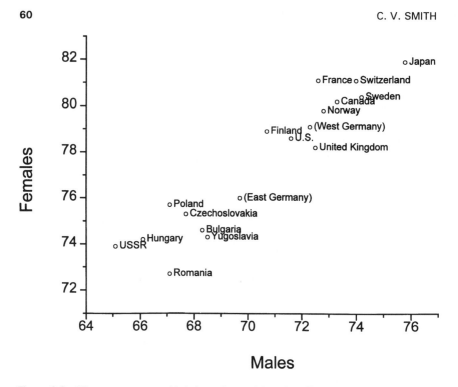

Figure 3-2 Life expectancy at birth in males and females. The data are adapted from Lopez (1990) for the period 1986–1988.

imposed on Eastern European countries over the last several decades has confused this process of "economic transition," leaving a more mixed pattern of public health issues. During the period between 1970 and 1985, heart disease mortality in men in Bulgaria increased by over 60%, by 35% in Poland, whereas the increase in Czechoslovakia was around 10%. In contrast, during this same period, heart disease mortality in Finnish males decreased approximately 25% and decreased in French males by over 30% (Smith & Marmot, 1991). Further, chronic heart disease and lung cancer have shown trends toward greater concentrations away from the economically privileged classes into the manual labor classes, apparently reflecting healthier diets and decreasing tobacco use in the former group.

Tobacco is the major cause of premature death in men of Eastern Europe and a large and growing problem in women. Tobacco is a primary problem in pediatric health because one of the primary objectives of any effort to improve the health of children is to enable them to grow into healthy adults. At present, tobacco use in the region involves about one-third of the population, although the use by males, up to 60% in Hungary, far exceeds use by women in most countries (Piha et al., 1993). Whereas tobacco use by Western European countries decreased by about 18% from 1974–1976 and 1984–1986, changes in Eastern Europe were characterized by increases or minor decreases.

Deaths due to lung cancer, which are largely attributable to cigarette smoking, increased dramatically in most industrialized countries over the first three decades following World War II. For example, between 1950 and 1980, lung cancer mortality rates more than tripled in U.S. males aged 65–74 years (Lopez, 1990). More recently, declines in age-specific lung cancer mortality rates in males have been observed in the United States and Western European countries, especially in younger age groups. Unfortunately, the lung cancer death rates in females in the United States and some Western European countries are continuing to increase, although encouraging trends are being observed in younger women. Attenuation and reversal of this growth in lung cancer deaths hopefully will be accomplished in Eastern European countries, although these goals appear to lie in the more distant future than in the Western countries.

Advertising by the multinational tobacco companies is sophisticated and effective and cannot be counterbalanced by fragmented and uncoordinated local governments whose primary priorities need to be the efforts to establish stable economies in the immediate future. Nevertheless, some countries have initiated innovative campaigns, such as the Columbus Project in the Czech Republic, whose theme is "Let's give tobacco back to America."

Limited data from Poland indicate that tobacco use in the 1980s decreased slowly in men, but increased in women. The greater incidence of tobacco use by younger women than by older women suggests a disturbing trend. Efforts by tobacco companies to target youthful markets in developing and transitional countries are unfortunate. The growth of market-driven economies in the transitional countries has attracted the attention of multinational companies. This attention may appear to offer a source of jobs and economic growth in countries in great need of such opportunities, but the longer range costs associated with health care problems may not be as apparent.

In contrast with the troublesome trends in increased tobacco use, health problems associated with alcohol consumption generally have remained more static (Lehto, 1993). Recent increases in mortality rates from chronic liver disease and cirrhosis of the liver in Romania and Bulgaria suggest the need for directed programs, designed to address the specific problems for each of these regions.

Unhealthy diets are major contributors to health problems in Eastern Europe (Gutzweiller, 1991; Nanda et al., 1993), often a result of general poverty, but also abetted by a lack of appreciation of the benefits that can be obtained by diet, exercise, and healthier lifestyle choices. Coronary heart disease tends to be lower in southern and eastern countries in Europe than in the English-speaking and the Scandinavian countries, apparently correlated with lower amounts of fats in the diets typical in the former regions. In contrast, death due to stroke is higher in Eastern Europe, probably because of poorer medical control of hypertension as well as lifestyle issues other than dietary fat, such as cigarette smoking, lack of exercise, and lower access to citrus fruits (Doyle, 1996).

GOVERNMENT, SOCIAL STRUCTURE, AND THE ROLE OF SCIENCE

Although economic development in Eastern Europe since World War II has been limited, what economic and industrial development has occurred often has been achieved with little regard for environmental health concerns or the principles of sustainable growth (Rose & Bloom, 1994). These oversights have had severe consequences for human health, as illustrated clearly by the lower life expectancies of people living in these countries than of citizens of the neighboring countries of Western Europe, or in Japan, Canada, or the United States (Figure 3-1). Over the preceding two decades the overall cancer rate declined in Western Europe, but doubled in Russia, where about half of the population uses drinking water that does not meet hygienic standards and 85% of the population lives in areas where pollution exceeds permissible levels (Rose & Bloom, 1994). Large segments of the populations of the other countries of Eastern Europe also live in regions that are excessively polluted. The newly formed governments and social structures are ill-prepared to identify, regulate, and correct the many factors that are contributing to the present environmental difficulties within their own borders.

Several factors contribute to the growing lack of scientific expertise in the countries of Eastern Europe and the newly independent states of the former Soviet Union. Whereas at one time many of the brightest students from these countries studied science in the educational systems that were supported strongly by the respective states, more recently the lack of support for education has made life exceedingly difficult for university professors, many of whom have to work at second jobs so that they can continue to teach (Koenig et al., 1996). The brightest students are not attracted to enter a profession in which, if they are successful, they will have to drive a cab at night so they can continue to teach and perhaps manage to conduct a little research. In addition, a coordinated, comprehensive, and systematic approach is needed to the issues of assessing the environmental hazards, correcting the problems, and improving human health in the region. However, the long history of warfare between countries and ethnic groups in the region will not make this cooperation easier to initiate and manage. Even within countries that have begun to address their own environmental problems, poor coordination of these efforts is a major problem.

One of the difficulties facing the formulation of optimal health policies is the difficulty in establishing a universally acceptable operational definition of a measure of benefit (Wolfson, 1994), which is essential in optimizing assessments of cost/benefit ratios. As regards the situation in Eastern Europe, the statistical base is confused, fragmented, and the preparation of any coordinated overview of the situation is difficult. Survival curves exist, but measures of health and productivity of the people while they are alive and the burdens of chronic disease are more difficult to quantitate. Such measures would give to people charged with the responsibility to establish policy the information needed to optimize their efforts. Such epidemiologic information also could be used as a very effective tool in setting priorities for basic health science research.

The hazards and risks associated with the use of and exposure to new chemicals, old chemicals placed into new uses, or chemicals presently in use, but with as yet unrecognized problems, exceed the capacity of toxicological evaluation, even in developed countries. In countries facing even harder choices for the allocation of much more limited resources, the greatest possible efficiency in risk evaluation and minimization processes is essential. In addition, Eastern European countries are witnessing the proliferation of small companies that, in the absence of adequate governmental oversight and regulation, essentially are left to their own devices with regard to establishing and maintaining healthy working environments and proper disposal of waste materials (Skowronska & Dawydzik, 1994). Human history suggests that this arrangement is unlikely to be sufficient. Nevertheless, these countries are unlikely to be able to commit large amounts of their limited resources to regulatory activities in the near or even in the foreseeable future. Hungary's budget for environmental health research in 1993 was only $50,000 (Balter, 1993). Only 2% of the total of $6 million funded by the U.S. National Institutes of Health for biomedical research in the region is directed at research in environmental health (Rose & Bloom, 1994). Improvements in environmental protection are more likely to arise from advances in the efficiency with which limited resources can be utilized and increasing emphasis on prevention rather than reliance on responses to crises. Knowledge gained through investigations of the fundamental mechanisms through which substances mediate undesired effects is a critical source of improvements in efficiency and thereby in environmental protection.

CONCLUSIONS

The preparation of this chapter left a haunting image of cultures that not so long ago were the most advanced in the world, at the very forefront of science, the arts, and progress in the human condition. The forces that led the countries of Eastern Europe to lose their positions of leadership in the world may be attributed to personal greed and lack of ethnic and religious tolerance. War is bad. No kidding. More relevant to the topic of the current discussion are the human and economic consequences of the failure of these countries to follow environmentally sound policies capable of sustaining continued growth and development. The route traveled by the countries of Eastern Europe over the last several decades looks uncomfortably like the choices that are facing the (presently) more developed countries. This is a road down which I do not wish to lead my children.

In addition, it is in the best interests of the people fortunate enough to have been born in developmentally advantaged countries (again, presently advantaged countries) to recognize the globally shared nature of the consequences of risk/ benefit analyses, particularly in view of the increasing magnitude of the effects of human activities on the global environment.

> We pray for children
>   whose nightmares come in the daytime,
>   who have no safe blanket to drag behind them,
>   who can't find any bread to steal,
>   and for those who will grab the hand of anybody kind enough to offer it.
> —Adapted from Ina J. Hughes (Clinton et al., 1990)

REFERENCES

Anonymous. 1995. Diptheria epidemic—New independent states of the former Soviet Union. *Morbid. Mortal. Weekly Rep.* 44:177–181.

Balter, M. 1993. Joining forces to probe environment–health links. *Science* 261:24–25.

Baverstock, K. F. 1993. Thyroid cancer in children in Belarus after Chernobyl. *World Health Statist. Q.* 46:157–176.

Cermerlic, E. Z., and Schaller, J. G. 1995. Human wrongs: A children's hospital destroyed. *J. Am. Med. Assoc.* 274:386.

Clinton, H. R., et al. 1990. *Children 1990. A report card, briefing book, and action primer.* Washington, DC: Children's Defense Fund.

Danzon, M., and Litinov, S. K. 1993. EUROHEALTH Programme. *World Health Stat. Q.* 46:153–157.

Dobrowolski, J. W., and Smyk, B. 1993. Environmental risk factors of cancer and their primary prevention. *J. Environ. Pathol. Toxicol. Oncol.* 12:55–57.

Doyle, R. 1996. Stroke mortality in men ages 35 to 74. *Sci. Am.* 274:26.

Dutkiewicz, T. 1991. The priority list of environmental chemical hazards in Poland. *Sci. Tot. Environ.* 101:153–158.

Falandysz, J. 1994. Some toxic and trace metals in big game hunted in the northern part of Poland. *Sci. Tot. Environ.* 141:59–73.

Gibbons, A. 1992. Researchers fret over neglect of 600 million patients. *Science* 256:1135.

Glosnicka, R., and Kunikowska, D. 1994. The epidemiological situation of *Salmonella entreitidis* in Poland. *Int. J. Food Microbiol.* 21:21–30.

Grant, J. P., and Nakajima, H. 1993. Hope for the future. *World Health* 46:3.

Gutzweiller, F. 1991. Health in central Europe. *Ann. Nutr. Metab.* 35(suppl. 1):64–68.

Hartvelt, F. 1993. The children's vaccine initiative. *World Health* 46(2):4–6.

Jenista, J. A. 1992. Disease in adopted children from Romania. *J. Am. Med. Assoc.* 268: 601–602.

Johnson, D. E., Miller, L. C., Iverson, S., Thomas, W., Franchino, B., Dole, K., Kiernan, M. T., Georgieff, M. K., and Hostetter, M. K. 1992. The health of children adopted from Romania. *J. Am. Med. Assoc.* 268:3446–3451.

Jones, I. M., Thomas, C. B., Tucker, C. L., Pleshanov, P., Vorobtsova, I., and Moore, D. H. 1995. Impact of age and environment on somatic mutation at the hprt gene of T lymphocytes in humans. *Mutat. Res.* 338:129–139.

Koenig, R., Milligan, S., Kahn, P., and Stone, R. 1996. After communism: Reinventing higher education. *Science* 271:695–701.

Krelowska-Kulas, M. 1995. Content of some metals in mean tissue of salt-water and fresh-water fish and in their products. *Nahrung* 39:166–172.

Kuchuk, A. A. 1994. Health problems of the population in different regions of the Ukraine. *Toxicol. Lett.* 72:213–217.

Kurtz, J. 1991. HIV infection and hepatitis B in adopted Romanian children. *Br. Med. J.* 302: 1399–1404.

Lehto, J. 1993. Alcohol consumption and related problems. *World Health Stat. Q.* 46:195–198.

Lopez, A. D. 1990. Who dies of what? A comparative analysis of mortality conditions in developed countries around 1987. *World Health Stat. Q.* 43:105–114.

Mackenzie, D. 1993. Europe's toxic waste. *World Health* 46(5):6–8.

Metelko, Z., Roglic, G., and Skrabalo, Z. 1992. Diabetes in time of armed conflict: The Croatian experience. *World Health Stat. Q.* 45:328–333.

Nanda, A., Nossikov, A., Prokhorskas, R., and Shabanah, M. H. A. 1993. Health in central and eastern countries of the WHO European Region: An overview. *World Health Stat. Q.* 46:158–165.

Norska-Borowka, I. 1994. Pediatric problems in Upper Silesia—Region of ecological disaster. *Toxicol. Lett.* 72:219–225.

Patrascu, I. V., and Dumitrescu, O. 1993. The epidemic of human immunodeficiency virus infection in Romanian children. *AIDS Res. Hum. Retrovir.* 9:99–104.

Piha, T., Besselink, E., and Lopez, A. D. 1993. Tobacco or health. *World Health Stat. Q.* 46:188–194.

Rooure, C., and Oblapenko, G. 1993. Communicable diseases in the CCEE/NIS. *World Health Stat. Q.* 46:177–187.

Rose, C. D., and Bloom, A. D. 1994. Human health and the environment in Eastern and Central Europe. *Environ. Health Perspect.* 102:696–698.

Schuler, D. 1992. Paediatric care in Hungary. *Arch. Dis. Child.* 67:1042–1045.

Skowronska, R., and Dawydzik, L. T. 1994. The sanitary inspection in the conditions of social and economic transformations in Poland. *Int. J. Occup. Med. Environ. Health.* 7:221–224.

Smith, G. D., and Marmot, M. G. 1991. Trends in mortality in Britain: 1920–1986. *Ann. Nutr. Metab.* 35(suppl. 1):53–63.

Solomons, N. W., Mazariegos, M., Brown, K. H., and Klasing, K. 1993. The underprivileged, developing country child: Environmental contamination and growth failure revisited. *Nutr. Rev.* 51:327–332.

Vienonen, M. A., and Wlodarczyk, W. C. 1993. Health care reforms on the European scene: Evolution, revolution or seesaw? *World Health Stat. Q.* 46: 166–169.

von Muhlendahl, K. E., and Otto, M. 1994. *Kinderarzt und Umwelt. Jahrbuch 1993/1994.* Munich: Alete Wissenschftlicher Dienst.

von Muhlendahl, K. E., Otto, M., and Spranger, J. 1995. Environmental aspects in pediatrics. *Eur. J. Pediatr.* 154:337–338.

Weinberg, A. D., Kripalani, S., McCarthy, P. L., and Schull, W. J. 1995. Caring for survivors of the Chernobyl disaster. *J. Am. Med. Assoc.* 274:408–412.

Wolfson, M. C. 1994. POHEM—A framework for understanding and modeling the health of human populations. *World Health Stat. Q.* 47:157–176.

Chapter 4

Environmental Chemicals
in Human Milk

Cheston M. Berlin, Jr., and Sam Kacew

The physiological process of breast-feeding has distinct advantages nutrition-ally, immunologically, and psychologically, and despite the presence of environ-mental toxins should be encouraged under most circumstances. There continues to be emphasis on promotion of breast-feeding by both professionals and lay persons. The incidence of breast-feeding in North America varied over the years from 25% (1970) to 56% (1981), and most recent studies show that about 54% of women discharged home are breast-feeding (Lawrence, 1989, 1994; Berlin, 1989; Buttar, 1994). Wilson et al. (1986) estimated that a figure of 70% of breast-feeding mothers in the United States was equivalent to approximately 2.5 million women. Even at 50%, in 1994 there were 4,000,000 new mothers in the United States, so more than 2 million infants went home receiving breast milk. A health goal established by a workshop sponsored by Human Health Services is that by the year 2000, 75% of newborns will be breast-fed and 50% will still be breast-fed by 6 months of age (U.S. Department of Health and Human Services, 1990). This numerical figure will no doubt rise as educational awareness of the benefits of lactation is promoted. Sufficient evidence exists that the benefits to the infant from human milk far outweigh the risks of toxic effects to chemicals.

The breast-feeding mother is subjected to exposure from environmental con-taminants, and these pollutants may be present in human milk. In the majority of cases it is still more beneficial to breast-feed despite the presence of milk contaminants, and the distinct advantages of breast-feeding are enhanced in con-ditions where mothers are nutritionally healthy. At present it should be stressed that human milk remains the best nutrient source for the healthy term infant (Redel & Shulman, 1994).

Breast-feeding is known to create a special psychological bond between infant and mother that ultimately leads to a socially healthier child (Newton & Newton, 1967). In addition, lactation enhances maternal postpartum recovery, and body weight returns to prepartum levels more rapidly. The distinct advantages of breast-feeding and human milk are widely appreciated, and it is recommended that the barriers that keep women from initiation or continuation of this physiological process be decreased (Lawrence, 1989). An all too often neglected factor in the initiation and acceptance of breast-feeding is the attitude of expectant fathers. Freed et al. (1992) demonstrated that knowledgeable fathers who believed that breast-feeding aided infant bonding and protected infants from disease promoted this physiological process. In contrast, misconceptions that breast-feeding interferes with sex and makes breasts ugly persuaded fathers to encourage the use of formula feeding (Jordan & Wall, 1990; Freed et al., 1992). A further misconception involves exercise during the postpartum period. Indeed, Dewey et al. (1994) found that women engaged in aerobic exercise 6–8 weeks postpartum displayed improved cardiovascular fitness with no adverse effect on lactation. It is evident that fathers also play an important role in the initiation and maintenance of the breast-feeding process (de Chateau et al., 1977).

The physiological process of breast-feeding plays a critical role in human growth and development. The use of bottles to feed nursing infants with milk formula in developing countries was found to enhance morbidity and mortality (Cunningham, 1979; Cunningham et al., 1991). In affluent nations inadequate knowledge on growth patterns between breast-fed versus bottle-fed infants resulted in inappropriate counseling against lactation (Whitehead & Paul, 1984). In an extensive study, Dewey et al. (1992) demonstrated that the growth pattern was equivalent between breast-fed and formula-fed infants from birth to the first 3 months. However, breast-fed infants gained significantly less weight from 3 to 12 months without any deleterious effects on nutrition, morbidity, activity level, or behavioral development. It is evident that breast-fed infants are leaner and do not display a faltering growth pattern; women should be encouraged to continue the lactational process beyond 3 months.

However, as pointed out by Krebs et al. (1994), maternal nutrient intake is an important factor in the growth pattern of the infant, and, for example, if a deficiency of zinc exists, there may be a delay in infant growth. Thus, in conditions of adequate maternal zinc concentrations lactation should be encouraged despite a decreased infant growth pattern. The growth pattern must be considered in light of chemical exposure and toxicity. Lipophilic toxicants including polychlorinated biphenyls (PCBs), hexachlorobenzene, organochlorine pesticides, etc. are present in human milk, stored in mammary fat, and adverse effects to both mother and nursing infant are related to concentration and release from fat (Borlakoglu & Dils, 1990; Peters et al., 1982; Dewailley et al., 1991; Quinby et al., 1965). Indeed, Falck et al. (1992) demonstrated a positive correlation between environmental contaminants such as PCBs or organochlorine residues stored in mammary adipose tissue and increase in maternal breast cancer frequency. Based on breast-fed infant growth patterns, a nursing infant possesses more body fat during the first 3 months and this baby would thus be more susceptible to lipophilic compound toxicity during this period. From 3 to 12 months a breast-fed infant is less fat than a bottle-fed child. Hence if a breast-fed infant is switched to a bottle at 3 months there would be an increase in the risk of toxicity to lipophilic agents between 3 and 12 months, as these infants possess more fat and thus potentially store more toxin. On the other hand, adverse reactions to water-soluble drugs including aspirin, lithium, cimetidine, etc. would be greater in the breast-fed infant compared to formula from 3 to 12 months as the infant is smaller and thus higher mammary drug concentrations would be attained (Kacew, 1993a, 1993b). The consequences of exposure to mammary chemicals or drugs need to be reassessed in light of differences in growth patterns between breast-fed versus formula-fed infants (Dewey et al., 1992). Growth or weight gain is frequently utilized as an index of toxicity in animals and humans (Kacew, 1990). With the knowledge that growth in breast-fed infants is generally less than with formula feeding, this parameter by itself may not be considered suitable to establish toxicity in the first year of life.

SUSCEPTIBILITY TO DISEASE

One important reason for the initiation and maintenance of breast-feeding is associated with the ability of human milk to reduce the incidence of morbidity and mortality in suckling infants compared to formula feeding (Wilson, 1983; Cunningham et al., 1991; May, 1988). Extensive studies clearly demonstrated that breast-feeding provides not only essential nutrition but also protection against infection and a variety of other immunological disorders (Welsh & May, 1979; Lawrence, 1989; Cunningham et al., 1991; May, 1994). A summary of the illnesses where it has been clearly demonstrated that breast-feeding provides protection to the nursing infant is given in Table 4-1. Although the evidence is overwhelming that there are distinct advantages in the breast-fed infant being protected from disease, one should be aware of several idiosyncrasies. Regardless of the climate or living circumstances, breast-feeding provides protection against infant respiratory disease manifestations including wheezing, bronchitis, and pneumonia (Wright et al., 1989; Cunningham et al., 1991). However, it should be noted that in farm workers in the grain or cotton industries, or in sewage plant employees, individuals are more readily exposed to gram-negative endotoxin bacteria (Lundholm et al., 1986; Dutkiewicz, 1986). A consequence of inhaling endotoxin is a marked rise in respiratory diseases including pneumonia. Of particular relevance was the findings of Gordon et al. (1993) that during lactation the maternal pulmonary system was more susceptible to endotoxin, resulting in greater lung injury to the nursing mother. It is of interest that during lactation there is also a greater susceptibility and increased pulmonary toxicity in ozone-exposed animals (Gunnison et al., 1992). Further morbidity from infectious diseases was higher in a polluted area where mothers were

Table 4-1   Importance of Lactation for Infant Health

Condition	Reference
Infant mortality	Madeley et al. (1986)
Sudden infant death syndrome	Cunningham et al. (1991)
Lower respiratory tract disease	Howie et al. (1990)
	Wright et al. (1989)
Food allergies	Gerrard et al. (1973)
Atopic dermatitis	Lucas et al. (1990)
Otitis media	Duncan et al. (1993)
	Saarinen (1982)
Chronic liver disease	Sveger and Udall (1985)
Immune system disorders	
Lymphomas	Davis et al. (1988)
Celiac disease	Greco et al. (1988)
Insulin-dependent diabetes mellitus	Mayer et al. (1988)
Inflammatory bowel disease	Whorwell et al. (1979)
Crohn's disease	Koletzko et al. (1989)
Antiparasite	
Entamoeba histolytica dysentery	Acosta-Altamirano et al. (1986)
Toxoplasma gondii CNS lesions	Mack and McLeod (1992)
Giardia lamblia diarrhea	Walterspiel et al. (1994)
Antiviral	
HIV-related mortality	Lederman (1992), Dunn et al. (1992)
Rotavirus diarrhea	Clemens et al. (1993)
Cytomegalovirus inclusion disease	May (1988)
Rubella virus	Welsh and May (1979), May (1988)
HTLV-1	Ando et al. (1989)
Antibacterial	
Cholera	Clemens et al. (1990)
Shigellosis	Ahmed et al. (1992)
Bacteremia and meningitis	Cochi et al. (1986)

exposed to inhaled particulate matter in combination with heavy metals (Tabacova & Vukow, 1992). Clearly, the living circumstances or environment is a factor to be considered in breast-feeding. As susceptibility to inhalation toxicity from endotoxin or ozone is enhanced during lactation, breast-feeding as a physiological process to protect against respiratory infections can only be encouraged through control of the environment. It should be noted that even during acute maternal infection the quantity and nutritional quality of human milk is not altered (Zavaleta et al., 1995).

It is generally accepted that breast-feeding delays the development of food allergies and atopic dermatitis (Lucas et al., 1990; Saarinen & Kajosaari, 1995). While Kramer (1988) suggested that breast-feeding provided protection against allergic diseases, these manifestations were also reported in exclusively breast-fed infants (Hattevig et al., 1987). However, it should be noted that the incidence of food allergies decreased dramatically in exclusively breast-fed infants (Gerrard et al., 1973). In recent studies, maternal diet was found to be a critical factor in the development of allergic manifestations in breast-fed infants. Ingestion of a maternal hypoallergenic diet devoid of cow's milk, eggs, and fish during lactation decreased the cumulative incidence and current prevalence of atopic dermatitis during the first 6 months of age with continuation to age 4 years (Chandra et al., 1989; Sigurs et al., 1992). Breast-feeding per se is effective in protecting against allergic disorders, provided the offending stimulus is not ingested by the mother for transmission to the infant via the milk.

The protection breast-fed infants receive against lymphoidal hypertrophy and lymphomas has been documented (Davis et al., 1988; Boat et al., 1975). This protective antineoplastic effect was found to extend to the mother where lactation for prolonged periods was correlated with a reduction in breast cancer (McTiernan & Thomas, 1986; Newcomb et al., 1994). Although Kvale and Heuch (1988) failed to demonstrate any association in a large cohort study, in the more extensive and recent findings of Newcomb et al. (1994) a positive inverse correlation between lactation and risk of breast cancer was noted in premenopausal women. However, no reduction in the risk of breast cancer was found in postmenopausal women with a history of lactation. Regardless of the fact that breast cancer occurs in less than one-quarter of all cases reported and that lactation provides a slight protective effect, Kelsey and John (1994) concluded that any factor that reduces the incidence of breast cancer should be encouraged. It should be noted that breast tissue serves as a storage depot for organochlorine pesticides, PCBs, etc. and that these environmental contaminants were implicated in the development of mammary tumors in animal studies (Berlin, 1989; Wasserman et al., 1976). Lactation in excessively environmentally exposed women would thus not be associated with a decreased risk of breast cancer.

The transmission of human immunodeficiency virus (HIV) during breast-feeding, from infected mothers to suckling infants, is well documented (Oxtoby, 1988; Gabiano et al., 1992; Dunn et al., 1992). Based on this knowledge the question arises as to the advantages of bottle-feeding in HIV-infected mothers. In an extensive study Lederman (1992) estimated that for exclusive breast-feeding, even in situations where mothers were HIV infected, there was a decrease in the estimated infant mortality rate, especially in population areas where HIV prevalence was low. Although lactation does not protect against HIV transmission, breast-feeding is clearly beneficial in reducing infant mortality, as breast milk is not the predominant route for HIV transmission (May, 1994). In a population where HIV infection is exceedingly high, bottle-feeding is preferable if safe formula feeding is available (Committee on Pediatric AIDS, 1995). In countries with no access to artificial formulas, benefits of breast-feeding substantially outweigh HIV transmission as one must consider infant mortality (Lederman, 1992). Viruses including cytomegalovirus (CMV) and hepatitis C virus (HCV) are transmitted to infants through breast milk without any apparent adverse effects (May, 1988; Zanetti et al., 1995). In contrast, human T-lymphotropic virus type I (HTLV-I) is transmitted mainly via breast milk to infants, and the longer the duration of lactation, the higher is the incidence of infectivity (Taka-hashi et al., 1991). HTLV-I produces adult T-cell leukemia and a myelopathy. Breast milk is known to contain a variety of antiviral factors to HIV, CMV, hepatitis B, and rotavirus, which reinforces the notion that breast-feeding pro-tects against infant mortality regardless of cause; this physiologic process should not be discontinued except under exceptional circumstances (Cunning-ham et al., 1991). The presence of antiparasitic and antibacterial factors in hu-man milk further demonstrates the importance of lactation in protecting the infant against disease (May, 1994).

GENERAL CONSIDERATIONS

The issue of excretion of maternal medications in milk has been well covered in many publications (Kacew, 1993a, 1993b, 1994; Berlin, 1992). This chapter dis-cusses issues surrounding the presence of environmental chemicals in human milk. It is not possible to mention all possible chemicals; there are literally thousands. As an example, Giroux et al. (1992) mention the Infotox database of the Commission de la santé et de la sécurité du travail (CSST), which contains 5736 compounds. Only 113 (2%) of these have evidence of transfer into human milk. For the remainder, data do not yet exist. In the absence of significant binding to maternal serum protein, compounds with a molecular weight below 200 are likely to be found in milk, especially if they are lipophilic (Rasmussen, 1971; Schanker, 1962). The ability to diffuse into milk appears directly related to lipid solubility (Schanker, 1962). The intensity of exposure to environmental substances will vary from country to country, and even in each country from region to region (industrial chemicals in urban areas, pesticides in rural areas, occupational exposures, and contaminated food chains are examples). This re-view attempts to focus on what is known of the excretion of environmental chemicals in human milk; animal studies are mentioned only if they provide insight into possible mechanisms of action of a particular compound. Many studies use units of concentration of parts per million (ppm, mg/kg) or parts per billion (ppb, μg/kg) to express concentration. In this chapter the units used are μg or ng/ml milk or per milligram milk fat to permit comparisons and to utilize units familiar to the reader.

The literature on this subject is vast; well over 1000 papers have been published, including the effects of environmental chemicals on milk production, milk composition, infant toxicity (mostly animal models), analytic techniques, and in vitro studies. An excellent source of details of most of these studies is the monograph by Jensen and Slorach (1991).

There are some important issues to consider in evaluating the effect of environmental chemicals on the entire process of lactation.

Background Levels

Many survey studies have shown widespread presence of chemicals by geography and social class of chemicals, although at very low concentrations (Savage et al., 1981; Pellizzari et al., 1982). Because many of these compounds are lipophilic, they are deposited in body fat. Lactation and pregnancy may represent the only route of elimination even after environmental exposure has ceased. For example, DDT was banned in Norway in 1970. Seven years later, nursing mothers still had detectable levels of DDT in their milk (Bakken & Seip, 1976; Brevik & Bjerk, 1978). Some of these chemicals are so ubiquitous that it may be impossible to find appropriate controls.

Analytical Technology

Analytical techniques have become extremely sophisticated and precise, both qualitatively and quantitatively. These have permitted the analyses of extremely low concentrations of chemicals and metabolites that previously were labeled as "below limit of detection." For example some dioxins can be measured in amounts as low as pg/mg milk fat (Weisglas-Kuperus et al., 1995). However, it should be noted that the study conditions, including analytic techniques, level of exposure, and sample collection, are not uniform. Thus, conclusions based on a single subject or small number of subjects may be misleading when attempts are made to translate small numbers of case reports into public health policy.

Exposure Conditions

It may not be possible to separate exposure during pregnancy from exposure during lactation. Most women excreting chemicals in their milk have been exposed for years to the chemicals, and placental transfer will also have occurred. Animal models may be useful here. Nielsen and Andersen (1995), using mice, showed that transfer of mercury (from methylmercury) during pregnancy is greater than that during lactation. Gladen et al. (1988) also came to the conclusion that transfer during pregnancy is quantitatively more important than during lactation. Except for isolated mass contamination events, exposure is generally over a period of years (perhaps the mother's lifetime) at very low levels. Infants exposed through breast-feeding should be studied over equally long periods, perhaps even into adult years, to ascertain as completely as possible any biological effects. Areas that need investigation include neurodevelopmental studies, physical growth, immune status, inducibility of mixed-function oxidases, and reproductive function. Second- and third-generation studies are also needed. These are expensive, laborious, and require considerable investigator dedication and, increasingly rare in current academic circles, longevity of investigators in the same institution or geographic site. Geographic moves of patients as well as investigators make such longitudinal studies very difficult.

In this chapter exposure of the nursing mother to environmental agents is discussed. Since there are thousands of possible agents, attention is focused on the more common exposures, especially those that may have resulted in clinical symptoms.

HALOGENATED HYDROCARBONS

DDT and Related Compounds

Perhaps the first report of an environmental chemical in human milk was by Laug et al. (1950, 1951). Techniques were not very precise by current technology standards. The average concentration of DDT was 130 ng/ml. More recent studies have also measured isomers, degradation products, and metabolites. These analyses have been done for other halogenated hydrocarbons (chiefly pesticides and fungicides). Almost all of the early studies (until 1980) showed detectable levels of DDT in nearly all women tested. Levels in Canada and the United States have remained at approximately 2 µg/ml from 1967 to 1982. A 1973 survey of seven United States cities showed a mean level of 0.17 µg/ml (Wilson et al., 1973). There is a suggestion of decline over a period as short as 4 years in Ontario (van Hove Holdrinet et al., 1977). Several other countries have reported a decline in levels of DDT over the years. In Western Australia DDT levels dropped from 3.6 to 0.8 µg/ml milk fat from 1974 to 1991 (Stevens et al., 1993); in Spain the decrease was from 65 to 0.4 µg/ml whole milk from 1979 to 1991 (Hernández et al., 1993); and in Sweden from 110 (1967) to 15 (1985) µg/ml (Westöö, 1974; Norén, 1988). Adverse effects on the human infant have not been described at these levels. In all these countries, decline was associated with a cessation in agricultural use. In an area of Mississippi contaminated by DDT, levels were 7.2 µg/ml, compared to 0.8 µg/ml in a noncontaminated part of the state (Barnett et al., 1979). With use of DDT for malaria control, levels were measured up to 15 ng/ml milk fat (Bouwman et al., 1992). It may take many years for levels to decline to undetectable levels because of the storage in body fat. The Western Australia study already mentioned did measure a decline in body fat from 2.75 µg/kg fat in 1970 to 0.72 µg/kg in 1991 (Stevens et al., 1993). A recent study showed levels in New York State of 0.56 µg/mg milk fat (pooled sample) (Schecter et al., 1989). Assuming each liter of human milk contains 36 g fat, this would be total daily exposure of 20 µg. It is not clear that such an exposure is hazardous. A recent report describes an inverse relationship between DDE (major metabolite of DDT) content of milk and duration of lactation (Gladen & Rogan, 1995). When the duration of breast-feeding was 7.5 months DDE levels of 2.5 µg/mg fat in milk were found, but with a nursing duration of only 3 months the concentration of DDE reached 12.5 or higher µg/mg milk fat. The authors postulated that this effect may be due to the estrogen activity (weak) of DDE and related isomers in reducing milk supply. Although DDT has been identified in breast milk, there were no apparent adverse effects noted in the suckling infant (Vuori & Tulinen, 1977).

Halogenated Biphenyls (PCBs and PBBs)

The biphenyl structure may have attached chlorine (polychlorinated or PCB) or bromine (polybrominated or PBB) in many ring positions, giving rise to a very large number of PCBs and PBBs. Skerfving et al. (1994) describe 209 congeners that were formerly used as insulation media in electrical systems, lubricants, and as paint additives. These compounds now are used only in closed components to prevent air, soil, and water contamination. One of the first episodes of environmental contamination occurred in Japan in 1968 (Yamaguchi et al., 1971). Rice cooking oil became contaminated with PCBs and some 1700 people became ill, with 20 deaths (Yusho disease). Ingestion of polychlorinated biphenyl-contaminated rice oil by nursing mothers was reported to produce low-birth-weight human infants, growth retardation, liver disease, abnormal skin pigmentation, and bone and tooth defects (Yamaguchi et al., 1971; Masuda et al., 1985). Since these compounds are taken up by body fat with subsequent excretion during both pregnancy and lactation, it is not possible to establish the risk only during lactation. These women obviously had a large body burden of PCBs. In Taiwan, congenital exposure to PCB contaminants (cooking oil) resulted in children who were shorter and lighter. They also had abnormalities of gingiva, nails, and teeth, and frequent pulmonary infections (Rogan & Ragan, 1994). Deficits of milestones were evident on Bayley, Stanford-Binet, or WISC Testing, but behavior (Rutter scale) seemed unaffected. In 1973 and 1974 there was widespread contamination of dairy cattle, pig, and chicken feed in Michigan with PBBs (Brilliant et al., 1978). Nearly all samples of human milk in the affected geographic area had detected levels of PBB between 0.05 to 1.0 µg/ml. A larger study of 1057 nursing women, also in Michigan, found PCBs in all samples in concentrations ranging from trace to 5.1 µg/ml (Wickizer & Brilliant, 1981). There appeared to be a correlation with the amount of fish consumed (Schwartz et al., 1983; Rogan et al., 1988). These levels may produce a total daily dose of 6 µg/kg/day, which may cause induction of hepatic microsomal enzymes (Rogan et al., 1986a, 1988). A decrease in PCB levels with duration of nursing and number of children nursed was reported (Rogan et al., 1986a; Hong et al., 1994) Widespread contamination of human milk has been noted in many other countries besides the United States (Bordet et al., 1993; Hernández et al., 1993; Duarte-Davidson et al., 1994). PCB (as well as dioxin) exposure has been shown to have adverse effects on immune function in animals (Tagaki et al., 1987). A recent Dutch study has shown lower monocyte and granulocyte counts in infants at 3 months of age, directly related to the amount of background exposure to PCBs and dioxins (Weisglas-Kuperus et al., 1995). In extensive studies in North Carolina, Rogan et al. (1986b) measured the levels of polychlorinated biphenyls in human milk and found an associated hypotonicity and hyporeflexia in nursing infants. In infants whose mothers were exposed to PCB and DDT, testing at 1 year showed decreased psychomotor scores with increased levels of maternal PCB burden (as measured in cord blood, placenta, maternal serum, and milk), but these scores were unaffected by breast-feeding (Gladen et al., 1988). In a recent study Huisman et al. (1995) found that higher concentrations of planar PCBs in breast milk correlated with a greater incidence of hypotonia. The long-term effects of such exposure are uncertain, but evidence indicates that these compounds exert neurotoxic effects on developing brain of infants (Huisman et al., 1995). Following the episodes of contamination in Japan and Michigan (USA) the Committee on Environmental Hazards of the American Academy of Pediatrics issued a statement that encouraged women to breast-feed unless there was a well-documented exposure to PCBs or PBBs. If the latter occurred, levels of the chemicals should be measured in milk (American Academy of Pediatrics, 1978).

Dioxins and Benzofurans

These compounds are frequently grouped together because they are by-products of the synthesis of 2,4,5-trichlorophenoxyacetic acid (Agent Orange herbicide) and 2,4,5-trichlorophenol (fungicide). The burning of PCBs and PBBs also yields chlorinated dioxins and furans, of which there are hundreds of possible congeners. In industrial countries the average content of human milk is about 2 pg/g milk fat. Levels in milk samples from South Vietnam in 1973 reached 4.27 pg/g milk, but declined to 1.11 pg/g 11 years later (Schecter et al., 1987, 1989). These data compare to 0.064 pg/g in North Vietnam (1984) and 1.04 pg/g in the United States. The effect of this amount on the infant is uncertain, although Koppe (1989) presents an intriguing theory that these compounds may be responsible for vitamin K deficiency bleeding (hemorrhagic disease of the newborn) in young breast-fed infants. Chloracne, seen after the industrial accident at Seveso, Italy, may have been caused by interference with the cytoskeleton of the skin by levels of dioxins 10 times "background" levels (13 pg/g) (Landers & Bunce, 1991). In a recent study, Pluim et al. (1993) found that maternal exposure to chlorinated dioxins and furans resulted in high blood thyroxine levels in infants at birth, but that at 11 weeks the thyroxine levels and the ratio of thyroxine to thyroxine-binding globulin became significantly elevated in neonates due to lactational exposure. Clearly human thyroid hormone regulation is affected by breast-milk dioxins. Further, high levels of dioxins in breast milk of nursing Norwegian women were associated with reduced neonatal neurological optimality (Huisman et al., 1995). As clinicians, one should be aware that the consequences of lactational exposure to toxicants can also be delayed. This is supported by the findings of Casey and Collie (1984) where a mother was exposed to 2,4-diphenoxyacetic acid (2,4-D) spray during pregnancy and lactation. Examination of the infant at 5 and 24 months of age revealed multiple malformations and severe mental retardation. Although the mammary tissue content of 2,4-D was not determined, prolonged maternal exposure with consequent transmission to the infant was suggested to result in the observed toxicity.

Organochlorine Insecticides

Aldrin, dieldrin, and endrin are naphthalene derivatives that were used as insecticides and have been identified in human milk (Egan et al., 1965; Riordan, 1987). The amounts in human milk has been decreasing over the years in countries where their use has been banned. A recent study from Western Australia demonstrated a steady decrease in dieldrin from 0.24 µg/g milk fat in 1974 to 0.05 µg/g in 1991 (Stevens et al., 1993). A similar decline in adipose tissue was also observed: from 0.19 to 0.04 µg/g extractable fat. All samples (milk and adipose tissue) did have detectable levels, illustrating the ubiquitous nature and long in vivo dwelling time of this pesticide. In Spain, only 12% of samples were positive in 1991 and showed a similar decline in levels from 1979/1982 to 1991 (Hernández et al., 1993). Similar data exist from the United States, but the size of the most recent sample is small ($n = 7$) and analyses were done on a pooled sample (Schecter et al., 1989). Heptachlor and chlordane are similar chemicals and are organochlorine insecticides used for control of the cotton boll weevil and termites. Heptachlor levels were between 35 and 90 ng/g milk fat (Savage et al., 1981; Bakken & Seip, 1976), while chlordane and metabolites levels were between 0.1 and 1.0 mg/ml (Miyazaki et al., 1980). However, it is surprising that manifestations of toxicity in suckling infants following maternal organochlorine exposure have not been reported. This should not be deemed evidence that the presence of organochlorine contaminants in breast milk lacks an effect on the infant, as these environmental toxicants induce mammary carcinoma and may act as cocarcinogens (Falck et al., 1992). It is well known that suckling infants of cigarette-smoking mothers are more prone to respiratory irritation and infections (Riordan, 1987). However, cigarette smoking in the presence of atmospheric pollutants exerted an additive toxic effect on the mother (Tabacova & Balabaeva, 1993). Conceivably, the presence of organochlorine compounds and nicotine in breast milk would increase infant toxicity, with the hydrocarbons acting as cocarcinogens. Hence in susceptible suckling infants these compounds may precipitate autoimmune diseases, lymphomas, etc. (Cunningham et al., 1991).

Lindane, or gamma-benzene hexachloride, is a pediculicide and scabicide. It has fallen into disfavor because of potentially significant percutaneous absorption causing central nervous system damage. In countries such as Nigeria where lindane is one isomer of the hexachlorocyclohexanes (HCH) used as insecticides, milk levels may be as high as 3 ng/ml (Atuma & Okor, 1987). In the United States analyses done 20 years ago showed levels of 36 pg/ml (Barnett et al., 1979), while in France no adverse effects were reported in infants suckling milk containing 48 pg/ml lindane (Luquet, 1975).

Hexachlorobenzene (HCB)

Hexachlorobenzene (HCB) is a fungicide but is also found as a contaminant in halogenated pesticides. It is found in nearly all samples of human milk at a level now of about 0.2 ng/ml. The human maternal ingestion of a fungicide, in hexachlorobenzene-treated wheat, resulted in chemical accumulation in breast milk. Suckling infants subsequently developed symptoms of a disease, pembe yara, and a condition of porphyria cutanea tarda (Cam & Nigogosyan, 1963; Peters et al., 1982), where the HCB levels reached as high as 2.8 µg/ml whole milk (Cripps et al., 1984). Levels in most other European countries have been between 2 and 10 ng/ml whole milk; some countries are higher, such as Spain with a mean level of 36 ng/ml whole milk (68% of samples positive) (Hernández et al., 1993). There has been a decrease in milk levels due to a cessation of use of HCB. With cessation of use in Western Australia, levels fell from 92 µg/ml milk in 1974 to 3.6 µg/ml in 1991 (Stevens et al., 1993). Levels of all halogenated hydrocarbons in infant formula, if present, are in amounts 1/16 and 1/90 of that in human milk (Jensen & Slorach, 1991).

Organophosphorous Insecticides

Exposure to organophosphate pesticides such as chlorpyrifos, malathion, etc. is worthy of mention, as these compounds have been identified in breast milk (Wilson, 1983; Chhabra et al., 1993). In a recent case, ingestion of chlorpyrifos by a 3-year-old child resulted in delayed polyneuropathy with transient bilateral vocal paralysis (Aiuto et al., 1993). Although lactation per se was not involved in this specific case, the importance lies in the fact that the manifestations of exposure did not occur until 1–3 weeks later. The fact that lactationally derived organophosphate pesticides alter suckling infant metabolism (Chhabra et al., 1993) and that toxicity may be delayed suggests that breast-feeding in severe exposure conditions should be minimized.

Solvents

The aromatic hydrocarbon toluene is utilized as a solvent or thinner in numerous industrial products including paints, glue, and resins. The lipophilic property of toluene is of interest in light of the physicochemical properties of breast tissue. Hersh et al. (1985) demonstrated that in children of approximately 4 years of age maternal exposure to toluene throughout pregnancy via glue sniffing produced embryopathy, mental deficiency, and postnatal growth delay. There is no doubt that in utero exposure to toluene was manifested in teratogenesis. However, as there was evidence of postnatal growth deficiency and toluene has a high affinity for fat, it is also possible that these children were exposed to this solvent via the mother's milk. This is supported by the finding that obstructive jaundice developed in lactating infants of mothers exposed to the dry-cleaning solvent perchloroethylene (Bagnell & Ellenberg, 1977). This child was breast-fed and developed jaundice at 5 weeks of age. The father was employed at a dry-cleaning plant and the mother visited him at lunch. Hyperbilirubinemia disappeared 8 days after breast-feeding was stopped. The infant's milk contained 100 ng/ml and her brother's blood contained 30 ng/ml of TCE. After 24 h of no exposure to the solvent, the milk level dropped to 30 ng/ml. The disappearance of jaundice with cessation of breast-feeding suggests a cause-and-effect relationship. Tetrachloroethylene is known to partition into breast milk, and various investigators have proposed an increased risk of cancer based on lactational exposure of infants to this chemical (Schreiber, 1993; Byczkowski & Fisher, 1995). It is well established that exposure to the organic solvents during pregnancy results in toxemia and anemia (Tabacova & Balabaeva, 1993). Unfortunately, infants born to these mothers were not followed during lactation. However, as these environmental chemicals accumulate in breast milk and produce metabolic maternal alterations, it is conceivable that solvent exposure may alter infant development. The release of organic solvents from breast-milk fat needs to be considered among solvent abusers in light of adverse consequent effects reported in children.

AFLATOXIN

Aflatoxins are produced by molds and are commonly present in foodstuffs, especially peanuts. They are thought to be carcinogens in animal models. In the tropics 28–37% of human milk samples contain aflatoxins in amounts as high as 56 ng/ml of milk (range 0.003–56) (Maxwell et al., 1989). Samples from France contained no detectable levels (Wild et al., 1987). Fusaric acid is a mycotoxin produced by the *Fusarium* species and is present in corn, wheat, barley, and other cereal grains. Recently, the dietary ingestion of fusaric acid by nursing dams resulted in lactational transfer of this mycotoxin to rat pups with suppression of pineal serotonin and tyrosine and decreased milk production (Porter et al.,1996). At present the effects of fusaric acid on human infants remain unknown, but fusaric acid may act to potentiate the adverse effects of other mycotoxins (Porter et al., 1993).

NICOTINE

Nicotine and its metabolite cotinine have been measured in human milk at levels of 12–48 ng/ml (Trundle & Skellern, 1983; Luck & Nau, 1985). The amount of nicotine is directly related to the number of cigarettes per day smoked (Luck & Nau, 1987; Anderson, 1982). The concentration of nicotine in milk is about three times that in maternal serum (Steldinger & Luck, 1988). The metabolite cotinine is also found in the urine of bottle-fed infants of mothers who smoke which is due to exposure to passive smoking (Labrecque et al., 1989; Schulte-Hobein et al., 1992). Mothers who smoke produce less milk and their infants have smaller weight gains than mothers who do not smoke (Hopkinson et al., 1992; Vio et al., 1991). It was found that in smokers the maternal concentration of plasma prolactin was reduced and that infants breathing smoke-filled air were more susceptible to respiratory irritation and infection (Riordan, 1987). It is important to note that cigarette smoke contains hundreds of other compounds, the toxicities of which have not be investigated in the nursing infant.

HEAVY METALS

Lactational exposure of human infants to metals is a concern and raises the issue of risk versus benefit in the maintenance of the breast-feeding process. Numerous studies exist on the effects of either prenatal or both during pregnancy and postnatal exposure to metals on developing infants, but few reports are available on the consequences of the presence of metals exclusively in breast milk on children. Virtually all metals are present in human milk; however, there is great variability associated with exposure of the mother (Kacew, 1994, 1996).

LEAD

In the United States the levels of lead in milk have varied according to year and geographical site (Knowles, 1974). Levels in Boston in 1979 averaged 1.7 ng/ml (Rabinowitz et al., 1985); in seven different U.S. cities Dillon et al. (1974) found levels averaging 26 ng/ml. In Mexico City, where there is considerable air pollution from gasoline lead, 54% of nursing mothers ($n = 35$) had no detectable levels; the remaining 46% had a mean level of 45 ng/ml (Namihara et al., 1993). Infants ingesting milk with this amount of lead would be exposed to a total daily dose 50% above the World Health Organization (WHO) permissible intake. It is uncertain what consequences of exposure to lead on the nursing infant over many months of lactation may result. There are widely known efforts of many public health organizations in the United States to minimize lead exposure to the preferred zero level. It is not surprising that upon examination of the source of infant lead intoxication, breast milk contained far less lead than either formula or environmental sources such as ceramic-leachable kitchenware, paint chips, etc. (Shannon & Graef, 1992; Rabinowitz et al., 1985). There is evidence to suggest a correlation between poor mental performance as evidenced by the Bayley Infants Assessment Test and increased lead levels (Needleman et al., 1983). These findings prompted Newman (1993) to recommend the promotion of breast-feeding, as human milk was a less suitable transmission vehicle for lead contamination compared to formula feeding. Further, it should be stressed that the best source for daily nutrition amongst infants is human milk (Redel & Shulman, 1994).

Although the contribution of exclusive mammary lead exposure that can be attributed to the observed toxic consequences remains undefined, there is evidence to suggest that the presence of lactational metal results in newborn toxicity (Sternowski & Wessolowski, 1985). Direct application of topical lead ointment on the breast was reported to produce central nervous system (CNS) toxicosis in the human infant (Dillon et al., 1974). The presence of lead in mammary tissue alters the nutritional value of milk as reflected by decreases in the essential elements copper, zinc, and iron (Bornschein et al., 1977), which are required for mammalian metabolism and CNS function. The absorption of calcium, iron, and vitamin D is affected by lead (Mahaffey, 1980). In conditions of diets deficient in essential elements, lead absorption and toxicity are enhanced in infants. As breast milk is a source of lead for suckling infants, it is conceivable that during iron-deficiency anemia or calcium-deficient dietary intake in mothers, the bioavailability of milk lead would be increased resulting in greater toxicity. Mahaffey (1980) suggested that lead exposure may interfere with maternal metabolic pathways, resulting in decreased utilization of nutrients in the diet, and thus an absence of nutritional components present in milk. This altered milk composition would consequently affect newborn development.

The beneficial effects of breast-feeding against immune system disorders are well established (Davis et al., 1988; Mayer et al., 1988; Kramer, 1988). The finding that lead interferes with the immune system (Pruett et al., 1993) indicates that the presence of this metal in milk may not confer protection in lead-exposed mothers. It should also be noted that in the presence of environmental toxicants and a condition of malnourishment, the immune system is further compromised as there is an increased sensitivity to viral infection (Porter et al., 1984). The effects of complex interactions between environment and diet on the suckling infant remain to be addressed. Although lactation was not the focal point, in an extensive study of approximately 250,000 births, Tabacova and Vukov (1992) found that excess of heavy metals including cadmium, lead, and chromium in combination with a deficiency of zinc and copper resulted in a significant incidence of spontaneous abortions, maternal toxemia, and congenital malformations. Particulate matter pollutants such as hydrocarbons, solvents, etc. in combination with heavy metal components led to rubella morbidity, a rise in congenital anomalies, and an increased frequency of complicated pregnancy (Tabacova & Balavaeva, 1993; Tabacova & Vukov, 1992). It is of interest that Mahaffey (1980) alluded to the combination of a deficiency in nutrients and excess mammary lead levels being associated with altered development in the suckling infant. In light of the fact that mammary tissue is exposed to a number of stimuli concurrently, a combination of factors rather than an individual environmental toxin should be utilized as determinants in the continuation of breast-feeding.

CADMIUM

In Toyama, Japan contamination of the river with cadmium effluent from industry resulted in metal accumulation in crops, ingestion of food prepared from contaminated crops, and poisoning of humans. The consequent condition termed "itai-itai" disease is characterized by bone pain, osteomalacia, and osteo-porosis (Murata et al., 1970). The effects of cadmium during pregnancy on maternal outcome and developing progeny, on pharmacokinetics, on toxicity in the neonate, and on the role of metallothionein have been reviewed by Bell (1984).

Administration of radioactive cadmium to lactating rats or mice revealed the presence of metal in mammary tissue and nursing infant organs (Lucis et al., 1972; Bell, 1984). Whelton et al. (1993a) demonstrated that in mice exposed to ^{109}Cd through 5 successive rounds of gestation there was a progressive accumu-lation of metal not only in the whole animal but a 14-fold increase in mammary tissue content. It is well known that cadmium interferes with the essential ele-ment zinc in placenta, and this may be reflected in the observed decrease in birth weight in metal-exposed animals (Kuhnert et al., 1988). Thus, it is con-ceivable that cadmium might induce a toxic response either through an action on mammary zinc content directly or through a transfer from mammary tissue, which then interferes with zinc in the nursing infant to decrease growth. It should be noted that in successive rounds of pregnancy cadmium content rises in mammary tissue but that the highest levels occur in kidney and liver, target organs for toxicity (Whelton et al., 1993a). It is of interest that in mice fed a diet resembling female itai-itai patients where there was a deficiency of calcium, phosphorous, iron, and vitamins A, B, and D, the accumulation of mammary cadmium was higher than that seen with a normal diet (Whelton et al., 1993b). Maternal malnutrition in cadmium-exposed lactating women would clearly enhance toxicity in suckling infants.

In humans, there is significant variation in the milk cadmium content, where Sternowski and Wessolowski (1985) found levels of 17 ng/ml (rural) and 25 ng/ml (urban) in the area of Hamburg, Germany, while Radisch et al. (1987) re-ported levels of 0.07 ng/ml (nonsmoking) and 0.16 µg/ml (smoking) in mothers in Berlin, Germany. The role of cadmium in human biology is not well defined. The relationship between smoking, the presence of nicotine in breast milk, and the consequent rise in infant respiratory infections and irritation is well known (Wilson, 1983; Riordan, 1987). Cigarette smoke is also the primary source of cadmium for humans (Lewis et al., 1972). Cadmium accumulation is approxi-mately sixfold greater in smokers compared to nonsmokers (Kuhnert et al., 1988). It has been suggested that a decrease in the placental zinc–cadmium ratio results in a lower newborn birth weight (Kacew, 1996). However, it is also conceivable that cadmium accumulates in breast milk with multiparity, is then transferred to the suckling infant, and subsequently mobilized in target organs (Chan & Cherian, 1993; Whelton et al., 1993a). Evidence clearly shows that maternal milk is a source of cadmium. In light of the observed decrease in birth weight in smoking mothers, the contribution of mammary cadmium toward the deposition of this metal to infant kidney, the site of toxicity, needs to be addressed.

MERCURY

Mercury levels in milk are usually low, <1 ng/ml (Pitkin et al., 1976). With environmental disasters (methylmercury contamination in Minamata, Japan) levels have been measured up to 50 ng/ml (Fujita & Takabatake, 1977). Of particular interest for this review is the Minamata Bay disaster, in which nursing mothers ingesting mercury-contaminated fish resulted in severe neurological disorders in human infants (Matsumoto et al., 1965). Takeuchi (1968) clearly demonstrated the effects of epidemic methylmercury exposure on fetal and new-born development in an extensive review. Industrial release of methylmercury into Minamata Bay followed by accumulation in edible fish and ingestion by lactating females resulted in the transfer of metal to the suckling human infant (Chang, 1984). Similarly, Amin-Zaki et al. (1974) demonstrated that ingestion by lactating mothers of homemade bread prepared from wheat treated with the fungicide methylmercury produced a significant rise in human infant metal levels to 200 ng/ml. In fish-eating populations in Canada, maternal ingestion of mercury-contaminated food during pregnancy and lactation resulted in abnormal muscle tone and reflexes in boys but not girls (McKeown-Eyssen et al., 1983). In a New Zealand study, Kjellstrom et al. (1989) reported developmental retardation in 4-year-old children of mothers eating mercury-contaminated fish during pregnancy and lactation. Although emphasis was placed on the consequences of prenatal exposure in the Canadian and New Zealand studies, the contribution of milk mercury to toxic outcome was neglected. It should be noted that in some patients with Minamata disease the neurological symptoms developed not at the time of exposure but years later (Igata, 1993). Further, Igata (1993) reported on a number of cases where children born in a mercury-contaminated area in Japan were healthy yet developed neuropathy in childhood. The contribution of breast-milk mercury to late-onset Minamata disease remains to be resolved. Mammary transfer of methylmercury to suckling infants has been reported to produce neurological lesions (Koos & Longo, 1976). This finding clearly indicates a positive correlation between exposure to high concentrations of metal in mammary tissue and toxicity in suckling infants.

Although the precise contribution of mammary-derived methylmercury to the observed adverse effects on neurologic and behavioral changes in suckling pups is not known (Chang, 1984), Nielsen and Andersen (1995) have shown that mercury transfer is quantitatively more important during pregnancy than lactation. The chemical form of the metal may also be important. In the guinea pig, mercury from methylmercury was taken up in higher concentrations in the kidney, liver, and brain than was mercury from mercuric chloride during both pregnancy and lactation (Yoshida et al., 1994). Khera and Tabacova (1973) found that postnatal exposure directly to newborns to this metal produced ocular defects. In contrast, there was a lack of an ocular effect in fetuses of prenatally exposed dams, suggesting that lactational methylmercury may in part contribute to the observed toxicity. It is well known that methylmercury is secreted more readily in the maternal colostrum, the period at which eye defects were reported, and crosses into the suckling infant. Mercury itself decreases the suckling response in human infants (Rohyans et al., 1984). Since milk contains essential nutrients for neurological and behavioral development, it is conceivable that less feeding would contribute to the mercury-induced nervous disorders as less nutritional supply is associated with delayed growth processes.

The protective role of breast-feeding against infectious diseases is well documented in noncontaminated mothers (Cunningham et al., 1991; Jason et al., 1984). Recently, Kosuda et al. (1994) demonstrated that methylmercury compromises the immune system by significantly reducing RT6-T lymphocytes in rats. Consequently, a renal autoimmune disease develops. It is conceivable that lactational mercury could serve as a potential source of metal, which ultimately leads to accumulation of metal in kidney and nephrotoxicity develops (Fowler, 1972) through action on the immune system. It is of interest that one of the manifestations of mercury poisoning in a human infant was hypersensitivity and allergic reactions, an index of immune system dysfunction (Rohyans et al., 1984). It is evident that methylmercury generates autoimmune antibodies, conceivably in maternal milk, and these may be transferred to the suckling infant. Indeed, Riordan (1987) suggested that if serum levels of maternal mercury are high or infants are symptomatic, breast-feeding should be terminated. Clearly, in exceptional circumstances the benefits derived from breast-feeding fail to outweigh the toxic consequences and this process should be terminated (Kacew, 1993a, 1993b).

SELENIUM

Selenium appears to be a necessary trace element for humans. It may function to protect cells against oxidation damage and perhaps protect against cancer (Alaejos & Romero, 1995). The enzyme glutathione peroxidase has selenium as a cofactor. Serum levels of this enzyme as well as selenium are low in newborn infants. In breast-fed infants, selenium content of milk varies with geographic locale (content of soil). Milk of women in high-selenium areas such as South Dakota will contain about 270 ng/ml of selenium. In low-selenium soil areas, the milk level may be 1/100 of that, 2.5 ng/ml (Keshan, China). Infants probably need about 10 μg/day, but this value is conjecture. Selenium supplementation of both pregnant and lactating women will increase selenium content of milk. Plasma selenium of breast-fed infants remains constant at about 43 ng/ml; infants fed formula drop their plasma selenium from 39 to 20 ng/ml at 4 months of age (Alaejos & Romero, 1995; Jochum et al., 1995). Even in areas of the world with high soil selenium, toxicity in nursing infants has not been described.

ANTIMONY

Antimony-containing compounds are used in the treatment of leishmaniasis. Elemental antimony can be transmitted to the infant through human milk.

Levels of 3.5 μg/ml may be achieved. This is higher than any of the metals discussed earlier. Even so, in an infant drinking 1 L milk per day, 3.5 mg antimony would be delivered. It is not established that this amount of elemental antimony can be absorbed from the gastrointestinal tract to produce detectable serum levels (Berman et al., 1989).

SUMMARY

With rare exception, and usually on the occasion of a large environmental contamination, chemical agents involved have not been significantly associated with effects in the nursing infant. What is now needed is large-scale studies, carried out over many years (especially important would be through puberty), utilizing current assaying techniques. In addition to studying biological influences, it will also be important to study psychological effects. Perhaps even the knowledge that human milk may be contaminated will result in concerns in the nursing mother that may interfere with breast-feeding (Hatcher, 1982).

REFERENCES

Acosta-Altamirano, G., Torres-Sanchez, E., and Meraz, E. 1986. Detection de anticuerpos de classe IgA dirigidos contra una lipopeptidofosfoglicana de *E. histolytica* en muestras de calostro humano. *Arch. Invest. Med.* 17:291.

Ahmed, F., Clemens, J. D., Rao, M. R., Sack, D. A., Khan, M. R., and Haque, E. 1992. Community-based evaluation of the effect of breast-feeding on the risk of microbiologically confirmed or clinically presumptive shigellosis in Bangladesh children. *Pediatrics* 90:406–411.

Aiuto, L. A., Pavlakis, S. G., and Boxer, R. A. 1993. Life-threatening organophosphate-induced delayed polyneuropathy in a child after accidental chlorpyrifos ingestion. *J. Pediatr.* 122:658–660.

Alaejos, M. S., and Romero, C. D. 1995. Selenium in human lactation. *Nutr. Rev.* 53:159–166.

American Academy of Pediatrics, Committee on Environmental Hazards. 1978. Statement Policy: PCBs in breast milk. *Pediatrics* 62:407.

Amin-Zaki, L., Elhassani, S., Majeed, M. A., Clarkson, T. W., Doherty, M. D., and Greenwood, M. R. 1974. Studies of infants postnatally exposed to methylmercury. *J. Pediatr.* 85:81–84.

Anderson, A. N. 1982. Suppressed prolactin but normal neurophysin levels in cigarette smoking breast feeding women. *Clin. Endocrinol.* 17:363–368.

Ando, Y., Saito, K., Nakano, S., Kakimoto, K., Furuki, K., Tanigawa, T., Hashimoto, H., Moriyama, I., Ichijo, M., and Toyama, T. 1989. Bottle-feeding can prevent transmission of HTLV-1 from mothers to babies. *J. Infect.* 19:25–29.

Atuma, S. S., and Okor, D. I. 1987. Organochlorine contaminants in human milk. *Acta Paediatr. Scand.* 76:365–366.

Bagnell, P. C., and Ellenberg, H. A. 1977. Obstructive jaundice due to a chlorinated hydrocarbon in breast milk. *Can. Med. Assoc. J.* 117:1047–1048.

Bakken, A. F., and Seip, M. 1976. Insecticides in human breast milk. *Acta Paediatr. Scand.* 65:535–539.

Barnett, R., D'Ercole, A. J., Cain, J. D., and Arthur, R. D. 1979. Organochlorine pesticide residues in human milk samples from women living in Northwest and Northeast Mississippi, 1973–75. *Pestic. Monit. J.* 13:47.

Bell, J. U. 1984. The toxicity of cadmium in the newborn. In *Toxicology and the newborn*, eds. S. Kacew & M. J. Reasor, pp. 199–216. Amsterdam: Elsevier.

Berlin, C. M., Jr. 1989. Drugs and chemicals: exposure of the nursing mother. *Pediatr. Clin. North. Am.* 36:1089–1097.

Berlin, C. M. 1992. The excretion of drugs and chemicals in human milk. In *Pediatric pharmacology*, 2nd ed., eds. S. Yaffe and J. Aranda, pp. 205–211. Philadelphia: W. B. Saunders.

Berman, J. D., Melby, P. C., and Neva, F. A. 1989. Concentration of Pentostam in human breast milk. *Trans. R. Soc. Trop. Med. Hyg.* 83:784–785.

Boat, T. F., Polmar, S. H., Whitman, V., Kleinerman, J. I., Stern, R. C., and Doershuk, C. F. 1975. Hyperreactivity to cow milk in young children with pulmonary hemosiderosis and cor pulmonale secondary to nasopharyngeal obstruction. *J. Pediatr.* 87:23–29.

Bordet, F., Mallet, J., Maurice, L., Borrel, S., and Venant, A. 1993. Organochlorine pesticide and PCB congener content of French human milk. *Bull. Environ. Contam. Toxicol.* 50:425–432.

Borlakoglu, J. T., and Dils, R. R. 1990. PCBs in human tissues. *Chem. Britain* 27:815–820.

Bornschein, R. L., Michaelson, I. A., Fox, D. A., and Loch, A. 1977. Evaluation of animal models used to study effects of lead on neurochemistry and behavior. *Symp. Biochemical Effects of Environmental Pollutants*, U.S. Environmental Protection Agency, Cincinnati, OH. Ann Arbor, MI: Ann Arbor Science.

Bouwman, H., Becker, P. J., Cooppan, R. M., and Reimcke, A. J. 1992. Transfer of DDT used in malaria control to infants via breast milk. *Bull. WHO* 70:241–250.

Brevik, E. M., and Bjerk, J. E. 1978. Organochlorine compounds in Norwegian human fat and milk. *Acta Pharmacol. Toxicol.* 43:59–63.

Brilliant, L. B., Van Amburg, G., Isbister, J., Humphrey, H., Wilcox, K., Eyster, J., Bloomer, A. W., and Price, H. 1978. Breast-milk monitoring to measure Michigan's contamination with polybrominated biphenyls. *Lancet* 2:643–646.

Buttar, H. S. 1994. Neonatal risks of drugs excreted in breast milk. *Can. Pharmaceut. J.* 127:14–19.

Byczkowski, J. Z., and Fisher, J. W. 1995. A computer program linking physiologically based pharmacokinetic model with cancer risk assessment for breast-fed infants. *Comput. Methods Programs Biomed.* 46:155–163.

Cam, C., and Nigogosyan, G. 1963. Acquired toxic porphyria cutanea tarda due to hexachlorobenzene. *J. Amer. Med. Assoc.* 183:88–91.

Casey, P. H., and Collie, W. R. 1984. Severe mental retardation and multiple congenital anomalies of uncertain cause after extreme parental exposure to 2,4-D. *J. Pediatr.* 104:313–315.

Chan, H. M., and Cherian, M. G. 1993. Mobilization of hepatic cadmium in pregnant rats. *Toxicol. Appl. Pharmacol.* 120:308–314.

Chandra, R. K., Puri, S., and Hamed, A. 1989. Influence of maternal diet during lactation and use of formula feeds on development of atopic eczema in high risk infants. *Br. Med. J.* 299:228–230.

Chang, L. W. 1984. Developmental toxicology of methylmercury. In *Toxicology and the newborn*, eds. S. Kacew and M. J. Reasor, pp. 173–197. Amsterdam: Elsevier.

Chhabra, S. K., Hashim, S., and Rao, A. R. 1993. Modulation of hepatic glutathione system of enzymes in suckling mouse pups exposed translactationally to malathion. *J. Appl. Toxicol.* 13:411–416.

Clemens, J., Sack, D., Harris, J., Khan, M. R., Chakraborty, J., Chowdhury, S., Rao, M. R., van Loon, F. P. L., Stanton, B. F., Yunis, M. D., Ali, M. D., Ansaruzzaman, M., Svennerholm, A. M., and Holmgren, J. 1990. Breast-feeding and the risk of severe cholera in rural Bangladeshi children. *Am. J. Epidemiol.* 131:400–411.

Clemens, J., Rao, M., Ahmed, F., Ward, R., Huda, S., Chakraborty, J., Yunis, M., Khan, M. R., Ali, M., Kay, B., van Loon, F., and Sack, D. 1993. Breast-feeding and the risk of life-threatening rotavirus diarrhea: Prevention or postponement? *Pediatrics* 92:680–685.

Cochi, S. L., Fleming, D. W., Hightower, A. W., Limpakarnjanarat, K., Facklam, R. R., Smith, J. D., Sikes, R. K., and Broome, C. V. 1986. Primary invasive Haemophilus influenzae type b disease: A population-based assessment of risk factors. *J. Pediatr.* 108:887–896.

Committee on Pediatric AIDS. 1995. Human milk, breastfeeding, and transmission of human immunodeficiency virus in the United States. *Pediatrics* 196:977–979.

Cripps, D. J., Peters, H. A., Gocmen, A., and Dogramici, I. 1984. Porphyria turcica due to hexachlorobenzene: A 20-30 year follow up study on 204 patients. *Br. J. Dermatol.* 111:413–422.

Cunningham, A. S. 1979. Morbidity in breast-fed and artificially fed infants. *J. Pediatr.* 95:685–689.

Cunningham, A. S., Jelliffe, D. B., and Jelliffe, E. F. P. 1991. Breast-feeding and health in the 1980s: A global epidemiologic view. *J. Pediatr.* 118:659–666.

Davis, M. K., Savitz, D. A., and Graubard, B. I. 1988. Infant feeding and childhood cancer. *Lancet* 2:365–368.

de Chateau, P., Holmberg, H., Jacobsson, K., and Winberg, J. 1977. A study of factors promoting and inhibiting lactation. *Dev. Med. Child. Neurol.* 19:575–584.

Dewailley, E., Weber, J.-P., Gingras, S., and Laliberté, C. 1991. Coplanar PCBs in human milk in the province of Québec, Canada: Are they more toxic than dioxin for breast fed infants? *Bull. Environ. Contamin. Toxicol.* 47:491–498.

Dewey, K. G., Heining, M. J., Nommsen, L. A., Peerson, J. M., and Lonnerdal, B. 1992. Growth of breast-fed and formula-fed infants from 0 to 18 months: The DARLING study. *Pediatrics* 89:1035–1041.

Dewey, K. G., Lovelady, C. A., Nommsen-Rivers, L. A., McCrory, M. A., and Lonnerdal, B. 1994. A randomized study of the effects of aerobic exercise by lactating women on breast-milk volume and composition. *N. Engl. J. Med.* 330:449–453.

Dillon, H. K., Wilson, D. J., and Schaffner, W. 1974. Lead concentrations in human milk. *Am. J. Dis. Child.* 128:491–492.

Duarte-Davidson, R., Wilson, S. C., and Jones, K. C. 1994. PCBs and other organochlorines in human tissue samples from the Welsh population: II—Milk. *Environ. Pollut.* 84:79–87.

Duncan, B., Ey, J., Holberg, C. J., Wright, A. L., Martinez, F. D., and Taussig, L. M. 1993. Exclusive breast-feeding for at least 4 months protects against otitis media. *Pediatrics* 91:867–872.

Dunn, D., Newell, M. L., Ades, A. E., and Peckham, C. S. 1992. Risk of human immuno-deficiency virus type 1 transmission through breast-feeding. *Lancet* 340:585–588.

Dutkiewicz, J. 1986. Microbial hazards in plants processing grain and herbs. *Am. J. Ind. Med.* 10:300–302.

Egan, H., Goulding, R., Roburn, J., and Tatton, J. 1965. Organo-chlorine pesticide residues in human fat and human milk. *Br. Med. J.* 2:66–69.

Falck, F., Jr., Ricci, A., Jr., Wolff, M. S., Godbold, J., and Deckers, P. 1992. Pesticides and polychlorinated biphenyl residues in human breast lipids and their relation to breast cancer. *Arch. Environ. Health.* 47:143–146.

Fowler, B. A. 1972. The morphologic effects of dieldrin and methylmercuric chloride on pars recta segments of rat kidney proximal tubules. *Am. J. Pathol.* 69:163–178.

Freed, G. L., Fraley, K., and Schanler, R. J. 1992. Attitudes of expectant fathers regarding breast-feeding. *Pediatrics* 90:224–227.

Fujita, M., and Takabatake, E. 1977. Mercury levels in human maternal and neonatal blood, hair and milk. *Bull. Environ. Contam. Toxicol.* 18:205–209.

Gabiano, C., Tovo, P. A., de Martino, M., Galli, L., Giaquinto, C., Loy, A., Schoeller, M. C., Giovannini, M., Ferranti, G., Rancilio, L., Caselli, D., Segni, G., Livadiotti, S., Conte, A., Rizzi, M., Viggiano, D., Mazza, A., Ferrazzin, A., Tozzi, A. E., and Cappello, N. 1992. Mother-to-child transmission of human immunodeficiency virus type 1: Risk of infection and correlates of transmission. *Pediatrics* 90:369–374.

Gerrard, J. W., Mackenzie, J. W. A., Goluboff, N., Garson, J. Z., and Maningas, C. W. 1973. Cow's milk allergy. *Acta Paediatr. Scand.* 234(suppl.):1–21.

Giroux, D., Lapointe, G., and Baril, M. 1992. Toxicological index and the presence in the work-place of chemical hazards for workers who breast-feed infants. *Am. Ind. Hyg. Assoc. J.* 53:471–474.

Gladen, B. C., and Rogan, W. J. 1995. DDE and shortened duration of lactation in a northern Mexican town. *Am. J. Public Health* 85:504–508.

88   C. M. BERLIN, JR., AND S. KACEW

Gladen, B. C., Rogan, W. J., Hardy, P., Thullen, J., Tingelstad, J., and Tully, M. R. 1988. Development after exposure to polychlorinated biphenyls and dichlorodiphenyldichloroethene transplacentally and through breast milk. *J. Pediatr.* 113:991–995.

Gordon, T., Weideman, P. A., and Gunnison, A. F. 1993. Increased pulmonary response to inhaled endotoxin in lactating rats. *Am. Rev. Respir. Dis.* 147:1100–1104.

Greco, L., Auricchio, S., Mayer, M., and Grimaldi, M. 1988. Case-control study on nutritional risk factors in celiac disease. *J. Pediatr. Gastroenterol. Nutr.* 7:395–399.

Gunnison, A. F., Weideman, P. A., and Sobo, M. 1992. Enhanced inflammatory response to acute ozone exposure in rats during pregnancy and lactation. *Fundam. Appl. Toxicol.* 19:607–612.

Hatcher, S. L. 1982. The psychological experience of nursing mothers upon learning of a toxic substance in their breast milk. *Psychiatry* 45:172–183.

Hattevig, G., Kjellman, B., and Bjorksten, B. 1987. Clinical symptoms and IgE responses to common food proteins and inhalants in the first 7 years of life. *Clin. Allergy* 17:571–578.

Hernández, L. M., Fernández, M. A., Hoyas, E. Gonzalez, M. J., and Garcia, J. F. 1993. Organochlorine insecticide and polychlorinated biphenyl residues in human breast milk in Madrid (Spain). *Bull. Environ. Contam. Toxicol.* 50:308–315.

Hersh, J. H., Podruch, P. E., Rogers, G., and Weisskopf, B. 1985. Toluene embryopathy. *J. Pediatr.* 106:922–927.

Hong, C. S., Xiao, J., Casey, A. C., Bush, B., Fitzgerald, E. F., and Hwang, S. A. 1994. Mono-*ortho*- and non-mono-*ortho*-substituted polychlorinated biphenyls in human milk from Mohawk and control women: Effects of maternal factors and previous lactation. *Arch. Environ. Contam. Toxicol.* 27:431–437.

Hopkinson, J. M., Schanler, R. J., Fraley, J. K., and Garza, C. 1992. Milk production by mothers of premature infants: Influence of cigarette smoking. *Pediatrics* 90:934–938.

Howie, P. W., Forsyth, J. S., Ogston, S. A., Clark, A., and Florey, C.V. 1990. Protective effect of breastfeeding against infection. *Br. Med. J.* 300:11–16.

Huisman, M., Koopman-Esseboom, C., Fidler, V., Hadders-Algra, M., van der Paauw, C. G., Tuinstra, L. G. M. T., Weisglas-Kuperus, N., Sauer, P. J. J., Touwen, B. C. L., and Boersma, E. R. 1995. Perinatal exposure to polychlorinated biphenyls and dioxins and its effect on neonatal neurological development. *Early Human Dev.* 41:111–127.

Igata, A. 1993. Epidemiological and clinical features of Minimata disease. *Environ. Res.* 63:157–169.

Jason, J., Nieburg, P., and Marks, J. S. 1984. Mortality and infectious disease associated with infant-feeding practices in developing countries. *Pediatrics* 74(suppl.):702–727.

Jensen, A. A., and Slorach, S. A. 1991. *Chemical contaminants in human milk.* Boca Raton, FL: CRC Press.

Jochum, F., Fuchs, A., Menzel, H., and Lombeck, I. 1995. Selenium in German infants fed breast milk or different formulas. *Acta Paediatr.* 84:859–862.

Jordan, P. L., and Wall, V. R. 1990. Breastfeeding and fathers: Illuminating the darker side. *Birth* 17:210–213.

Kacew, S. 1990. Developmental aspects of pediatric pharmacology and toxicology. In *Drug toxicity and metabolism in pediatrics,* ed. S. Kacew, pp. 1–13. Boca Raton, FL: CRC Press.

Kacew, S. 1993a. Adverse effects of drugs and chemicals in breast milk on the nursing infant. *J. Clin. Pharmacol.* 33:213–221.

Kacew, S. 1993b. Neonatal toxicology. In *General and applied toxicology,* eds. B. Ballantyne, T. Marrs, and P. Turner, vol. 2, pp. 1047–1068. London: Macmillan Press.

Kacew, S. 1994. Current issues in lactation: Advantages, environment, silicone. *Biomed. Environ. Sci.* 7:307–319.

Kacew, S. 1996. Mammary heavy metal content: Contribution of lactational exposure to toxicity in suckling infants. In *Toxicology of metals,* vol. II, ed. L. W. Chang, pp. 1129–1137. Boca Raton, FL: Lewis Press.

Kelsey, J. L., and John, E. M. 1994. Lactation and the risk of breast cancer. *N. Engl. J. Med.* 330:136–137.

Khera, K. S., and Tabacova, S. A. 1973. Effects of methylmercuric chloride on the progeny of mice and rats treated before or during gestation. *Food Cosmet. Toxicol.* 11:245–254.

Kjellstrom, T., Kennedy, P., Wallis, S., Stewart, A., Friberg, L., Lind, B., Wutherspoon, P., and Mantall, C. 1989. Physical and Mental Development of Children with Prenatal Exposure to Mercury from Fish. Stage 2. Interviews and Psychological Tests at Age 6, p. 112. Solna: National Swedish Environmental Board, report no. 3642.

Knowles, J. A. 1974. Breast milk: A source of more than nutrition for the neonate. *Clin. Toxicol.* 7:69–82.

Koletzko, S., Sherman, P., Corey, M., Griffiths, A., and Smith, C. 1989. Role of infant feeding practices in development of Crohn's disease in childhood. *Br. Med. J.* 298:1617–1618.

Koos, B. J., and Longo, L. D. 1976. Mercury toxicity in the pregnant woman, fetus, and newborn infant: A review. *Am. J. Obstet. Gynecol.* 126:390–409.

Koppe, J. 1989. Dioxins and furans in the mother and possible effects on the fetus and newborn breast-fed baby. *Acta Paediatr. Scand. (Suppl.)* 360:146–153.

Kosuda, L. L., Hosseinzadeh, H., Greiner, D. L., and Bigazzi, P. E. 1994. Mercury-induced renal autoimmunity: Experimental manipulations of "susceptible" and "resistant" rats. *J. Toxicol. Environ. Health* 42:303–321.

Kramer, M. S. 1988. Does breast feeding help protect against atopic disease? Biology, methodology and a golden jubilee of controversy. *J. Pediatr.* 112:181–190.

Krebs, N. F., Reidinger, C. J., Robertson, A. D., and Hambidge, K. M. 1994. Growth and intakes of energy and zinc in infants fed human milk. *J. Pediatr.* 124:32–39.

Kuhnert, D. R., Kuhnert, P. M., and Zarlingo, T. J. 1988. Associations between placental cadmium and zinc and age and parity in pregnant women who smoke. *Obstet. Gynecol.* 71:67–70.

Kvale, G., and Heuch, I. 1988. Lactation and cancer risk: Is there a relation specific to breast cancer? *J. Epidemiol. Community Health* 42:30–37.

Labrecque, M., Marcoux, S., Weber, J.-P., Fabia, J., and Ferron, L. 1989. Feeding and urine cotinine values in babies whose mothers smoke. *Pediatrics* 83:93–97.

Landers, J. P., and Bunce, N. J. 1991. The Ah receptor and the mechanism of dioxin toxicity. *Biochem. J.* 276:273–287.

Laug, E. P., Prickett, C. S., and Kunze, F. M. 1950. Survey analyses of human milk and fat for DDT content. *Fed. Proc.* 9:294.

Laug, E. P., Kunze, F. M., and Prickett, C. S. 1951. Occurrence of DDT in human fat and milk. *Arch. Ind. Hyg.* 3:245–246.

Lawrence, R. A. 1989. Breastfeeding and medical disease. *Med. Clin. North. Am.* 73:583–603.

Lawrence, R. A. 1994. *Breastfeeding: A guide for the medical profession*, 4th ed. St. Louis, MO: C. V. Mosby.

Lederman, S. A. 1992. Estimating infant mortality from human immunodeficiency virus and other causes in breast-feeding and bottle-feeding populations. *Pediatrics* 89:290–296.

Lewis, G. P., Jusko, W. J., Coughlin, L. L., and Hartz, S. 1972. Contribution of cigarette smoking to cadmium accumulation in man. *Lancet* 2:291–292.

Lucas, A., Brooke, O. G., Morley, R., Cole, J. T., and Bamford, M. F. 1990. Early diet of preterm infants and development of allergic or atopic disease: Randomized prospective study. *Br. Med. J.* 300:837–840.

Lucis, O. J., Lucis, R., and Shaikh, Z. A. 1972. Cadmium and zinc in pregnancy and lactation. *Arch. Environ. Health* 25:14–22.

Luck, W., and Nau, H. 1985. Nicotine and cotinine concentrations in serum and urine of infants exposed via passive smoking or milk from smoking mother. *J. Pediatr.* 107:816–820.

Luck, W., and Nau, H. 1987. Nicotine and cotinine concentrations in the milk of smoking mothers: Influence of cigarette consumption and diurnal variation. *Eur. J. Pediatr.* 146:21–26.

Lundholm, M., Palmgren, U., and Malmberg, P. 1986. Exposure to endotoxin in the farm environment. *Am. J. Indust. Med.* 10:314–315.

Luquet, F. M. 1975. Pollution of human milk in France with organochlorine insecticide residues. *Pathol. Biol.* 23:45–49.

Mack, D. G., and McLeod, R. 1992. Human *Toxoplasma gondii*-specific secretory immunoglobin A reduces *T. gondii* infection of enterocytes in vitro. *J. Clin. Invest.* 90:2585–2592.

Madeley, R. J., Hull, D., and Holland, T. 1986. Prevention of postneonatal mortality. *Arch. Dis. Child.* 61:459–463.

Mahaffey, K. R. 1980. Nutrient–lead interactions. In *Lead toxicity*, eds. R. L. Singhal and J. A. Thomas, pp. 425–460. Baltimore, MD: Urban and Schwarzenberg.

Masuda, Y., Kuroki, H., Haraguchi, K., and Nagayama, J. 1985. PCB and PCDF congeners in the blood and tissues of yusho and yu-cheng patients. *Environ. Health Perspect.* 59:53–58.

Matsumoto, M., Koya, G., and Takeuchi, T. 1965. Fetal Minimata disease. *J. Neuropathol. Exp. Neurol.* 24:563–574..

Maxwell, S. M., Apeagyei, F., de Vries, H. R., Mwamuit, D. O., and Hendrick, R. G. 1989. Aflatoxins in breast milk, neonatal cord blood and sera of pregnant women. *J. Toxicol. Toxin Rev.* 8:19–29.

May, J. T. 1988. Microbial contaminants and antimicrobial properties of human milk. *Microbiol. Sci.* 5:42–46.

May, J. T. 1994. Antimicrobial factors and microbial contaminants in human milk: Recent studies. *J. Paediatr. Child Health* 30:470–475.

Mayer, E. J., Hamman, R. F., Gay, E. C., Lezotte, D. C., Savitz, D. A., and Klingensmith, G. J. 1988. Reduced risk of IDDM among breastfed children. *Diabetes* 37:1625–1632.

McKeown-Eyssen, G. E., Ruedy, J., and Neims, A. 1983. Methylmercury exposure in northern Quebec. II. Neurological findings in children. *Am. J. Epidemiol.* 118:470–479.

McTiernan, A., and Thomas, D. B. 1986. Evidence for a protective effect of lactation on risk of breast cancer in young women: Results from a case-control study. *Am. J. Epidemiol.* 124:353–358.

Miyazaki, T., Akiyama, K., Kaneko, S., Horii, S., and Yamagishi, T. 1980. Chlordane residues in human milk. *Bull. Environ. Contam. Toxicol.* 25:518–523.

Murata, I., Hirono, T., Saeki, Y., and Nakaga, W. S. 1970. Cadmium enteropathy, renal osteomalacia ("itai-itai" disease in Japan). *Bull. Soc. Int. Clin.* 1:34–42.

Namihira, D., Saldivar, L., Pustilinik, Carreon, G. J., and Salinas, M. E. 1993. Lead in human blood and milk from nursing women living near a smelter in Mexico City. *J. Toxicol. Environ. Health* 38:225–232.

Needleman, H., Bellinger, D., Leviton, A., Rabinowitz, M., and Nichols, M. 1983. Umbilical cord blood lead levels and neuropsychological performance at 12 months of age. *Pediatr. Res.* 17:179A.

Newcomb, P. A., Storer, B. F., Longnecker, M. P., Mittendorf, R., Greenberg, E. R., Clapp, R. W., Burke, K. P., Willett, W. C., and MacMahon, B. 1994. Lactation and reduced risk of premenopausal breast cancer. *N. Engl. J. Med.* 330:81–87.

Newman, J. 1993. Would breast-feeding decrease risks of lead intoxication? *Pediatrics* 90:131.

Newton, N., and Newton, M. 1967. Psychologic aspects of lactation. *N. Engl. J. Med.* 277:1179–1188.

Nielsen, J. B., and Andersen, O. 1995. A comparison of the lactational and transplacental deposition of mercury in offspring from methylmercury-exposed mice. Effect of seleno-L-methionine. *Toxicol. Lett.* 76:165–171.

Norén, K. 1988. Changes in the levels of organochlorine pesticides, polychlorinated biphenyls, dibenzo-*p*-dioxins and dibenzofurans in human milk from Stockholm, 1972–85. *Chemosphere* 17:39–49.

Oxtoby, M. J. 1988. Human immunodeficiency virus and other viruses in human milk: Placing the issues in broader perspective. *Pediatr. Infect. Dis. J.* 7:825–835.

Pellizzari, E. D., Hartwell, T. D., Harris, B. S. H ., Waddell, R. D., Whitaker, D. A., and Erickson, M. D. 1982. Purgeable organic compounds in mothers milk. *Bull. Environ. Contamin. Toxicol.* 28:322–328.

Peters, H. A., Gocmen, A., Cripps, D. J., Bryan, G. T., and Dogramaci, I. 1982. Epidemiology of hexachlorobenzene-induced porphyria in Turkey. *Arch. Neurol.* 39:744–749.

Pitkin, R. M., Bahns, J. A., Filer, I. J., and Reynolds, W. A. 1976. Mercury in human maternal and cord blood, placenta, and milk. *Proc. Soc. Exp. Biol. Med.* 151:565–567.

Pluim, H. J., de Vijlder, J. M. M., Olie, K., Kok, J. H., Vulsma, T., van Tijn, D. A., van der Slikke, J. W., and Koppe, J. G. 1993. Effects of pre- and postnatal exposure to chlorinated dioxins and furans on human neonatal thyroid hormone concentrations. *Environ. Health Perspect.* 101:504-508.

Porter, J. K., Voss, K. A., Chamberlin, W. J., Bacon, C. W., and Norred, W. P. 1993. Neurotransmitters in rats fed fumonisin B1. *Proc. Soc. Exp. Biol. Med.* 202:360–364.

Porter, J. K., Wray, E. M., Rimando, A. M., Stancel, P. C., Bacon, C. W., and Voss, K. A. 1996. Lactational passage of fusaric acid from the feed of nursing dams to the neonatal rat and effects on pineal neurochemistry in the F1 and F2 generations at weaning. *J. Toxicol. Environ. Health* 49:161–175.

Porter, W. P., Hinsdill, R., Fairbrother, A., Olson, L. J., Jaeger, J., Yuill, T., Bisgaard, S., Hunter, W. G., and Nolan, K. 1984. Toxicant-disease-environment interactions associated with suppression of immune system, growth, and reproduction. *Science* 224:1014–1017.

Pruett, S. B., Ensley, D. K., and Crittenden, P. I. 1993. The role of chemical-induced stress responses in immunosuppression: A review of quantitative associations and cause-effect relationships between chemical-induced stress responses and immunosuppression. *J. Toxicol. Environ. Health* 39:163–192.

Quinby, G. E., Armstrong, J. F., and Durham, W. F. 1965. DDT in human milk. *Nature* 207:726–728.

Rabinowitz, M., Leviton, A., and Needleman, H. 1985. Lead in milk and infant blood: A dose-response model. *Arch. Environ. Health* 40:283–286.

Radisch, B., Luck W., and Nau, H. 1987. Cadmium concentrations in milk and blood of smoking mothers. *Toxicol. Lett.* 36:147–152.

Rasmussen, F. 1971. Excretion of drugs by milk. In *Handbook of experimental pharmacology*, vol. 28, *Concepts of biochemical pharmacology*, part I, eds. B. B. Brodie and J. R. Gillette, pp. 390–402. New York: Springer-Verlag.

Redel, C. A., and Shulman, R. J. 1994. Controversies in the composition of infant formulas. *Pediatr. Clin. North Am.* 41:909–924.

Riordan, J. 1987. Drugs excreted in human breast milk. In *Problems in pediatric drug therapy*, 2nd ed., eds. L. A. Pagliaro and A. M. Pagliaro, pp. 195–258. Hamilton, IL: Drug Intelligence Publications

Rogan, W. J., and Ragan, N. B. 1994. Chemical contaminants, pharmacokinetics, and the lactating mother. *Environ. Health Perspect.* 102(suppl. II):89–95.

Rogan, W. J., Gladen, B. C., McKinney, J. D., Carreras, N., Hardy, P., Thullen, J. D., Tingelstad, J., and Tully, M. 1986a. Polychlorinated biphenyls (PCBs) and dichlorodiphenyldichloroethene (DDE) in human milk: Effects of maternal factors and previous lactation. *Am. J. Public Health* 76:172–177.

Rogan, W. J., Gladen, B. C., McKinney, J. D., Carreras, N., Hardy, P., Thullen, J., Tingelstad, J., and Tully, M. 1986b. Neonatal effects of transplacental exposure to PCBs and DDE. *J. Pediatr.* 109:335–341.

Rogan, W. J., Gladen, B. C., Hung, K.-L., Koong, S., Shih, L., Taylor, J. S., Wu, Y., Yang, D., Ragan, N. B., and Hsu, C. C. 1988. Congenital poisoning by polychlorinated biphenyls and their contaminants in Taiwan. *Science* 241:334–336.

Rohyans, J., Walson, P. D., Wood, G. A., and MacDonald, W. A. 1984. Mercury toxicity following merthiolate ear irrigations. *J. Pediatr.* 104:311–313.

Saarinen, U. M. 1982. Prolonged breastfeeding as prophylaxis for recurrent otitis media. *Acta Paediatr. Scand.* 71:567–571.

Saarinen, U. M., and Kajosaari, M. 1995. Breastfeeding as prophylaxis against atopic disease: Prospective follow-up study until 17 years old. *Lancet* 346:1065–1069.

Savage, E. P., Keefe, T. J., Tessari, J. D., Wheeler, H. W., Applehans, F. N., Goes, E. A., and Ford, S. A. 1981. National study of chlorinated hydrocarbon insecticide residues in human milk, USA I. Geographic distribution of dieldrin, heptachlor, heptachlor epoxide, chlordane, oxychlordane, and Mirex. *Am. J. Epidemiol.* 113:413–422.

Schanker, L. S. 1962. Passage of drugs across body membranes. *Pharmacol. Rev.* 14:501–530.

Schecter, A., Ryan, J. J., and Constable, J. D. 1987. Polychlorinated dibenzo-*p*-dioxin and polychlorinated dibenzofuran levels in human breast milk from Vietnam compared with cow's milk and human milk from the North American continent. *Chemosphere* 16:2003–2016.

Schecter, A., Füerst, P., Kruger, C., Meemken, H.-A., Groebel, W., and Constable, J. D. 1989. Levels of polychlorinated dibenzofurans, dibenzodioxins, PCBs, DDT and DDE, hexachlorobenzene, dieldrin, hexachlorocyclohexanes and oxychlordane in human breast milk from the United States, Thailand, Vietnam and Germany. *Chemosphere* 18:445–454.

Schreiber, J. S. 1993. Predicted infant exposure to tetrachloroethene in human breastmilk. *Risk Anal.* 13:515–524.

Schulte-Hobein, B., Schwartz-Bickenbach, D., Abt, S., Plum, C., and Nau, H. 1992. Cigarette smoke exposure and development of infants throughout the first year of life: Influence of passive smoking and nursing on cotinine levels in breast milk and infant's urine. *Acta Paediatr.* 81:550–557.

Schwartz, P. M., Jacobson, S. W., Fein, G., Jacobson, J. L., and Price, H. A. 1983. Lake Michigan fish consumption as a source of polychlorinated biphenyls in human cord serum, maternal serum, and milk. *Am. J. Public Health* 73:293–296.

Shannon, M. W., and Graef, J. W. 1992. Lead intoxication in infancy. *Pediatrics* 89:87–90.

Sigurs, N., Hattevig, G., and Kjellman, B. 1992. Maternal avoidance of eggs, cow's milk, and fish during lactation: Effect on allergic manifestations, skin-prick tests, and specific IgE antibodies in children at age 4 years. *Pediatrics* 89:735–739.

Skerfving, S., Svensson, B.-G., Asplund, L., and Hagmar, L. 1994. Exposure to mixtures and congeners of polychlorinated biphenyls. *Clin. Chem.* 40:1409–1415.

Steldinger, R., and Luck, W. 1988. Half lives of nicotine in milk of smoking mothers: Implications for nursing. *J. Perinat. Med.* 16:261–262.

Sternowski, J. H., and Wessolowski, R. 1985. Lead and cadmium in breast milk. Higher levels in urban vs rural mothers during the 1st 3 months of lactation. *Arch. Toxicol.* 57:41–45.

Stevens, M. F., Ebell, G. F., and Psaila-Savona, P. 1993. Organochlorine pesticides in Western Australian nursing mothers. *Med. J. Aust.* 58:238–241.

Sveger, T., and Udall, J. N., Jr. 1985. Breastfeeding, alpha-1-antitrypsin deficiency, and liver disease? *J. Am. Med. Assoc.* 254:3036–3037.

Tabacova, S., and Balabaeva, L. 1993. Environmental pollutants in relation to complications of pregnancy. *Environ. Health Perspect.* 101(suppl. 2):27–31.

Tabacova, S., and Vukov, M. 1992. Issues of human exposure to agents causing developmental toxicity. *Congenital Anom.* 32(suppl.):S21-S30.

Tagaki, Y., Aburada, S., Otake, T., and Ikegami, N. 1987. Effect of polychlorinated biphenyls (PCBs) accumulated in the dam's body on mouse filial immunocompetence. *Arch. Environ. Contamm. Toxicol.* 16:375–381.

Takehashi, K., Takazaki, T., Oki, T., Kawakami, K., Yashiki, S., Fujioshi, T., Usuku, K., Mueller, N., Osame, M., Miyata, K., Nagata, Y., and Sonoda, S. 1991. Inhibitory effect of maternal antibody on mother-to-child transmission of human T-lymphotropic virus type-1. *Int. J. Cancer* 49:673–677.

Takauchi, T. 1968. Pathology of Minamata disease. In *Minimata disease*, ed. M. Kutsune, pp. 141–228. Kunamoto, Japan: Kunamoto University.

Trundle, J. I., and Skellern, G. G. 1983. Gas chromatographic determination of nicotine in human breast milk. *J. Clin. Hosp. Pharm.* 8:289–293.

U.S. Department of Health and Human Services. 1990. Healthy People 2000. Publication no. (PHS) 91-50213, Washington, DC.

van Hove Holdrinet, M., Braun, H. E., Frank R., Stopps, G. J., Smout, M. S., and McWade, J. W. 1977. Organochlorine residues in human adipose tissue and milk from Ontario residents, 1969–1974. *Can. J. Public Health* 68:74–80.

Vio, F., Salazar, G., and Infante, C. 1991. Smoking during pregnancy and lactation and its effects on breast-milk volume. *Am J. Clin. Nutrit.* 54:1011–1016.

Vuori, E., and Tylinen, H. 1977. The occurrence and origin of DDT in human milk. *Acta Paediatr. Scand.* 66:761–765.

Walterspiel, J. N., Morrow, A. L., Guerrero, L., Ruiz-Palacios, G., and Pickering, L. K. 1994. Secretory anti-*Giardia lamblia* antibodies in human milk: Protective effect against diarrhea. *Pediatrics* 93:28–31.

Wasserman, M., Nogueira, D. P., Tomatis, L., Mirra, A. P., Shibata, H., Arie, G., Cucos, S., and Wasserman, D. 1976. Organochlorine compounds in neoplastic and adjacent apparently normal breast tissue. *Bull. Environ. Contam. Toxicol.* 15:478–484.

Weisglas-Kuperus, S. T. C. J., Koopman-Esseboom, C., van der Zwan, C. W., de Ridder, M. A. J., Beishuizen, A., Hooijkaas, H., and Sauer, P. J. J. 1995. Immunologic effects of background prenatal and postnatal exposure to dioxins and polychlorinated biphenyls in Dutch infants. *Pediatr. Res.* 38:404–410.

Welsh, J. K., and May, J. T. 1979. Anti-infective properties of breast milk. *J. Pediatr.* 94:1–9.

Westöö, G. 1974. Changes in the levels of environmental pollutants (Hg, DDT, Dieldrin, PCB) in some Swedish foods. *Ambio* 3:79–83.

Whelton, B. D., Moretti, E. S., Peterson, D. P., and Bhattacharyya, M. H. 1993a. Cadmium-109 metabolism in mice. I. Organ retention in mice fed a nutritionally sufficient diet during successive rounds of gestation and lactation. *J. Toxicol. Environ. Health* 38:115–129.

Whelton, B. D., Moretti, E. S., Peterson, D. P., and Bhattacharyya, M. H. 1993b. Cadmium-109 retention in mice. II. Organ retention in mice fed a nutritionally deficient diet during successive rounds of gestation and lactation. *J. Toxicol. Environ. Health* 38:131–145.

Whitehead, R. G., and Paul, A. A. 1984. Growth charts and the assessment of infant feeding practices in the Western world and in developing countries. *Early Human Dev.* 9:187–207.

Whorwell, P. J., Holdstock, G., Whorwell, G. M., and Wright, R. 1979. Bottle feeding, early gastroenteritis, and inflammatory bowel disease. *Br. Med. J.* 1:382.

Wickizer, T. M., and Brilliant, L. B. 1981. Testing for polychlorinated biphenyls in human milk. *Pediatrics* 68:411–415.

Wild, C. P., Pionneau, F. A., Montesano, R., Mutiro, C. F., and Chetsanga, C. J. 1987. Aflatoxin detected in human breast milk by immunoassay. *Int. J. Cancer* 40:328–333.

Wilson, D. J., Locker, D. J, Ritzen, C. A., Watson, J. T., and Schaffner, W. 1973. DDT concentrations in human milk. *Am. J. Dis. Child* 125:814–817.

Wilson, J. T. 1983. Contamination of human milk by drugs and chemicals. *Nutr. Health* 2:191–201.

Wilson, J. T., Hinson, J. L., Brown, R. D., and Smith, I. J. 1986. A comprehensive assessment of drugs and chemical toxins excreted in breast milk. In *Human lactation*, vol. 2, eds. M. Hamosh and A. D. Goldman, pp. 395–423. New York: Plenum.

Wright, A. L., Holberg, C. J., Martinez, F. D., Morgan, W. J., and Taussig, L. M. 1989. Breast-feeding and lower respiratory tract illness in the first year of life. *Br. Med. J.* 299:946–949.

Yamaguchi, A., Yoshimura, T., and Kuratsune, M. 1971. A survey of pregnant women having consumed rice oil contaminated with chlorobiphenyls and their babies. *Fukuoka Acta Med.* 62:117–122.

Yoshida, M., Watanabe, C., Satoh, H., Kishimoto, T., and Yamamura, Y. 1994. Milk transfer and tissue uptake of mercury in suckling offspring after exposure of lactating maternal guinea pigs to inorganic or methylmercury. *Arch. Toxicol.* 68:174–178.

Zanetti, A. R., Tanzi, E., Paccagnini, S., Principi, N., Pizzocolo, G., Caccamo, M. L., D'Amico, E., Cambie, G., and Vecchi, L. 1995. Mother-to-infant transmission of hepatitis C virus. *Lancet* 345:289–291.

Zavaleta, N., Lanata, C., Butron, B., Peerson, J. M., Brown, K. H., and Lonnerdal, B. 1995. Effect of acute maternal infection on quantity and composition of breast milk. *Am. J. Clin. Nutr.* 62:559–563.

Chapter 5

Transport of Organic Chemicals to Breast Milk: Tetrachloroethene Case Study

Judith S. Schreiber

Exposure to and consequent storage of fat-soluble environmental contaminants by human beings has been documented repeatedly (Rogan & Gladen, 1991; Rogan et al., 1986a, 1986b; Jensen, 1983, 1991; Savage et al., 1981; Wallace et al., 1989). Contaminants such as 1,1-bis(*p*-chlorophenyl)-2,2,2-trichloroethane (DDT) and its metabolites, mirex, polychlorinated biphenyls (PCBs), heptachlor, aldrin/dieldrin, chlordane, and hexachlorobenzene (HCB) are environmentally and bio-logically persistent. These compounds are preferentially stored in fat tissue but have been found in virtually all body tissues. Adverse health effects have been associated with exposure to these chemicals, although not specifically as a result of breast-milk exposure.

One potential source of exposure to fat-soluble and other environmental contaminants is human breast milk. Mature human milk contains about 4% lipid, which supplies about 40% of the total calories in milk (Lawrence, 1989). Human milk fat can originate from the diet, from adipose tissue stores, and from de novo synthesis in the mammary gland. The relative proportions contributed by these three sources have been estimated to be 30%, 60% and 10%, respectively (see pages 100–103). The impact of dietary composition, nutritional status, stage of lactation, maternal weight loss, and related demands on adipose tissue will affect this distribution, but the changes due to these factors are not well characterized (Hachey et al., 1987; *Nutrition Reviews*, 1989). The lipid composition of milk remains remarkably constant despite a wide variation of dietary and other factors.

Infant formulas have been developed to mimic the composition of breast milk, with limited success.

A woman's exposure to chemicals prior to and during lactation can influence the composition of her breast milk. Chemicals that are readily metabolized and are eliminated from the mother's body in the urine and feces are unlikely to be secreted in significant quantities in breast milk unless the exposure is very recent and of sufficiently high magnitude (WHO, 1986). Conversely, chemicals that are poorly metabolized and slowly eliminated from the mother's body are more likely to be detected in breast milk. The degree to which a particular chemical enters the mammary gland depends on a variety of factors. In general, transfer of chemicals from blood plasma to milk is enhanced by high lipid solubility, low relative molecular mass, and weak binding to plasma proteins.

Although there is little doubt that environmental chemicals are present in the vast majority of human milk samples, the significance of the presence of these contaminants on the well-being of the mother and infant is not known. Studies on breast milk are usually undertaken for purposes of monitoring exposure to and storage of persistent environmental chemicals, and do not include monitoring of disease incidence or adverse effects in the children.

Concerns surrounding contamination of breast milk and its evaluation are not new. In 1908, C. B. Reed, an assistant professor of obstetrics at Northwestern University Medical School wrote:

> The removal of the nursing babe from the breast may be required for any one of a great variety of conditions which affect the mother or child or both. It may be demanded of a consequence of changes in the milk itself through chemical action, or through the presence in the milk of foreign substances which have been taken in by the mother, or which occur as the result of changes in the environment . . . the transmission of medical substances through the milk of the mother to the babe is a subject of more than usual interest and more than usual obscurity. Many cases have been reported with the object of putting the phenomenon on a real scientific basis, but owning to the prevalence of tradition and the absence of thorough and accurate methods in collecting the data, the observations are for the most part quite valueless. (p. 514)

When available, studies examining the impact of breast-milk contaminants on infant health generally consist of clinical case reports identifying acute adverse effects on the infant, such as intoxication and epidemiologic studies seeking to determine adverse effects that may not be manifested until many years later. However, whether the available epidemiologic and clinical case-study reports represent sufficient evidence to conclude that environmental contaminant exposure via ingestion of breast milk can result in adverse health impact on the infant has not been determined. Since many environmental contaminants can be transported across the placenta to the fetus, it is often difficult to assess separately the effect of the prenatal exposure and that of subsequent exposure of the infant to the contaminant via breast milk. Moreover, if the infant and mother are exposed to similar environmental contaminants (e.g., lead in the home), breast milk may not be the infant's chief source of postnatal exposure.

Information on the toxicity of chemicals in humans is often sparse and limited to high-dose or mixed chemical exposures, usually as a result of accidents or occupational exposure. Toxicity information on chemicals is often derived through testing on laboratory animals. In controlled experiments, animals are exposed to known amounts of the pure chemical and effects on the animals and their offspring are observed. From these studies, valuable information on the toxicity of the chemicals can be obtained. An acceptable daily intake (ADI) or reference dose (RfD) for humans may be calculated for noncarcinogenic effects from the results of appropriate animal studies. Chemicals that have been shown to be carcinogenic in animal studies are considered potential carcinogens in humans. Human cancer potency estimates can be calculated based on the findings of animal studies. The infant breast-milk dose can then be compared to doses associated with increased cancer risk, and can be compared to the RfD and/or ADI.

The beneficial aspects of human breast milk have been evaluated qualitatively, but rarely on a quantitative basis. Breast milk provides an economical source of high-quality nutrition for infants and young babies, in addition to providing immunological protection. The degree to which breast-feeding provides advantages to the infant will be quantitatively assessed for a variety of endpoints including infant mortality and morbidity rates.

COMPOSITION OF BREAST MILK

Over the past two decades, much has been learned about the composition of human milk. A thorough presentation of the components of human milk can be found in several reviews on the subject (Lawrence, 1989; Casey & Hambridge, 1983; Raiha, 1989; Hurley & Lonnerdal, 1988).

Mammals are the only animals able to produce milk to nurture their offspring. In general, the nutrient composition of the milk of each species is best suited to the growth needs of the offspring. As an evolutionary development, milk composition and volume represent a compromise between the maintenance of optimal nutrient intake for the offspring and the need to minimize the nutritional drain on the lactating female. Nutrient levels in milk, therefore, are likely to meet, but not greatly exceed, the requirements of the offspring (Raiha, 1989).

The synthesis of human milk is a complex process that is anticipated during pregnancy as the maternal body undergoes anatomical and physiological changes in the breast and elsewhere. When lactation begins there is a marked increase in metabolic rate, a redistribution of the blood supply, and an increased demand for nutrients (Lawrence, 1989; *Nutrition Reviews*, 1989). Human milk is a complex mixture of over 200 constituents including lipids, carbohydrates, proteins, vitamins, and minerals that is designed to provide the infant with a nutritionally complete food. The increased mammary blood flow, increased cardiac output, and milk secretion are suckling dependent and regulated by pituitary hormones.

   The composition of human milk varies with the stage of lactation, the time of day, the sampling time during a given day's feeding, maternal nutrition, and individual variation (Lawrence, 1989). Colostrum (produced in the first 5 days after birth), transitional milk (days 6–10), and mature milk have notable differences in their composition. Although infant formula composition attempts to mimic the composition of human milk, there are differences in the concentrations of a variety of minerals, fats, and other components in formula, human milk, and cow's milk. The differences lead, in some cases, to measurable differences in certain biochemical parameters in the infant such as serum amino acid, mineral and cholesterol concentrations, body fat composition, and urinary composition.

   The major components of milk are fats, carbohydrates, proteins, amino acids, and immunoglobulins, vitamins, minerals, and trace elements. Fats are the most variable component of human milk and their origin in breast milk is complex. Because of the lipid solubility of tetrachloroethene and other contaminants, the fat content of breast milk plays an important role in the infant dose.

   The fats that make up breast milk are derived from the mother's dietary intake, her adipose tissue stores, and mammary synthesis. The fats provide most of the caloric energy of milk and can vary considerably during the course of a single lactation period and over time. The fats in infant formula are usually provided by vegetable oils such as corn, sesame, soybean, coconut, or palm oils.

   Newborn infants, whose primary energy source during gestation has been glucose, use up their stores of liver glycogen as a glucose source within 24 h after birth. Glucose and fatty acids from the infant's diet and body fat are then used. Free fatty acids are released from triglycerides by a lipase present in human milk. The switch to fatty acids as an energy source requires a source of carnitine to transport them, as newborns do not have a fully developed ability to synthesize carnitine. They obtain it naturally from breast milk or as a supplement to infant formula. Bottle-fed infants whose formula does not contain carnitine often have impaired ability to use fatty acids and to regulate body heat and blood lipid levels (Guthrie & Bagby, 1989).

   Mature human milk contains about 4% lipid, which supplies about 40% of the total calories in milk. Individual milk lipid percentage varies widely, however, from about 1% to more than 10%. The lipids in milk are comprised primarily (98%) of triglycerides (TG) contained in a membrane-bound fat droplet called the milk fat globule. Other lipid constituents of milk include cholesterol (0.5%), phospholipids (1%), vitamin A, vitamin E, vitamin D, and a large number of minor lipids. The lipids contain mostly long-chain fatty acids (>C-16), which originate from the diet and from mobilized adipose tissue stores.

   It has long been thought that if caloric demands were met by an appropriate dietary intake of the mother, the lipids in milk fat would arise primarily from current dietary intake (Garza et al, 1987; Gallenberg & Vodicnik, 1989). However, interpretation of recent studies detailed later suggest that a substantial proportion of the fats in milk arises from previously deposited adipose tissue (fat depots) even if caloric demands are met. In addition to lipids, adipose tissue contains chemicals to which the mother has previously been exposed (by dermal contact, ingestion and inhalation) that are stored in body fat. Fat-soluble contaminants present in adipose tissue are mobilized as the adipose tissue is utilized for milk production.

The relationship of the contribution of adipose lipids and dietary lipids in the synthesis of breast milk is complex. The period of lactation puts nutritional demands on the maternal body as ingested nutrients and adipose stores are used for the production of milk. The initiation of lactation results in marked alterations in the partitioning of nutrients and in maternal metabolism to meet the demands of milk production. The partitioning of nutrients to various body tissues involves a dynamic equilibrium between two types of regulation: homeostasis and homeorhesis (Bauman & Currie, 1980).

After the ingestion of nutrients, the body attempts to maintain physiological equilibrium in an effort to maintain constant internal body conditions. For example, the postprandial period (after consumption of a meal) results in changes in concentrations of blood metabolites and an increase in the ratio of insulin to glucagon. These changes in turn result in a greater uptake of glucose by liver (for glycogen synthesis) and by adipose tissue (for lipid synthesis). In the postabsorptive period (1–2 h after consumption of a meal) when the uptake of nutrients has diminished, the ratio of insulin to glucagon decreases enough to cause glycogen and lipid stores to be mobilized. Therefore, with each cycle of feeding, there is a shift in the overall energy metabolism from a catabolic state (breaking down stored lipids in adipose tissue to provide energy) to an anabolic state (storing triglycerides as lipids in adipose tissue to provide needed energy later, after the meal). A constant supply of nutrients to peripheral body tissues is maintained by promotion of short-term storage of nutrients following a meal and mobilization of these stored nutrients during the postabsorptive period (homeostasis).

The second type of control in partitioning nutrients is homeorhesis. Bauman and Currie (1980) define homeorhesis as the coordinated changes in metabolism of body tissues necessary to support a physiological state, in this case lactation. In lactation, the metabolic functions of many maternal organs are altered in order to supply the mammary gland with sufficient nutrients for the synthesis of milk. One of the major changes occurs in adipose tissue, where the uptake of nutrients for synthesis of storage lipids is decreased and lipid reserves are mobilized instead. The processes of lipid synthesis (lipogenesis) include de novo synthesis of fatty acids from acetate in plasma, the uptake of fatty acids from circulating lipoproteins from the diet, stimulated by the enzyme lipoprotein lipase, and the esterification of these fatty acids into triglycerides (the storage form of lipid). The deposition and mobilization of fatty acids go on all the time, even when the mass of adipose tissue is not changing. Indeed, the triglycerides in adipose tissue are also in a dynamic state, even when the total mass of triglyceride is stable (Steinberg, 1985).

Lipolysis or mobilization of adipose tissue to free fatty acids takes place concurrently. Lipolysis leads to an outflow of free fatty acids and glycerol from adipose tissue by hydrolysis. The glycerol provides the energy needed by the peripheral tissues while the free fatty acids circulate in the blood plasma transported by albumin (Rosell & Belfrage, 1979). The regulation of homeostasis and homeorhesis is coordinated, and the factors that directly affect one process tend to cause a reciprocal change in the other (Bauman & Currie, 1980).

Human milk fat can originate from the diet, from adipose tissue stores, and from de novo synthesis in the mammary gland. The relative proportions contributed by these 3 sources has been estimated to be 30%, 60%, and 10%, respectively, although the impact of dietary composition, nutritional status, stage of lactation, maternal weight loss, and related demands on adipose tissue has not been well defined (*Nutrition Reviews*, 1989). Even though the overall percentage of lipids in breast milk changes, the specific lipids that comprise breast milk remain remarkably constant despite a wide variation of dietary and other factors.

Triglycerides (TG) come from the plasma and are also synthesized from intracellular glucose oxidized via the pentose pathway. In lactating humans (and other mammals), chylomicrons, very-low-density lipoproteins (VLDL) and low-density lipoproteins (LDL), carry dietary TG to the mammary gland. Free fatty acids and 2-monoacylglycerols are produced from the conversion of TG by lipoprotein lipase in the mammary gland. These free fatty acids and 2-monoacylglycerols are then resynthesized in the endoplasmic reticulum to produce new milk TG.

Blood circulation in adipose tissue has been poorly understood until recently, when advances in the morphology and physiology of adipose tissue have shed some light on the relationship of the mobilization of fat and its transport in the blood to various organs. Adipose tissue has an extensive capillary network around each adipocyte. The rich vascular supply to each adipocyte allows transcapillary transport of the products of lipolysis, that is, fatty acids and glycerol, to the general circulation. Although less perfused than more metabolically active tissue such as skeletal muscle (50–70 ml/min/100 g), adipose tissue blood flow at maximal vasodilation is significant at 25–30 ml/min/100 g. Resting blood flow in adipose tissue is usually between 7 and 12 ml/min/100 g based on experimental data. Values of 2–3 ml/min/100 g have been reported based on data from the washout of inert gases (Rosell & Belfrage, 1979).

The following studies have helped elucidate the origins of milk fat lipids. In a study of 3 healthy lactating women (Hachey et al., 1987), food supplements containing TG of deuterium-labeled palmitic (C16:0), oleic (C18:1), and linoleic (C18:2) acids were ingested for 3 days prior to analysis of milk and plasma in a voluntary restricted diet study. The women had been successfully lactating for periods of 7 to 28 weeks at the start of the study period. Peak enrichment of the labeled TG in plasma occurred within 2–4 h (much of it found in the chylomicrons and VLDL), and in milk within 8–10 h after consumption.

The observed 6-h delay between the peak enrichment in plasma and milk was due to the time differential between delivery of the tracer to the breast, lipolysis of chylomicron and VLDL TG in the mammary gland capillary bed, resynthesis into TG, extrusion of the milk fat globule, and expression of the milk from the breast. About 97% of the deuterium-labeled dietary TG found in the milk (approximately 5.1% of the total tracer administered) was secreted into the milk within 24 h. Almost 60% of the fatty acids incorporated into milk fat within 72 h (the duration of the study) was not derived from the isotopically labeled dietary fatty acids, indicating extensive mobilization of fatty acids from adipose tissue stores, where the lipid-soluble toxicants are stored. Notably, these women were not intentionally dieting or involved in weight-reduction programs; the diet provided in the study contained the number of calories normally ingested by these women during lactation. Contaminant mobilization was not studied.

Wilson et al. (1988) studied the sources of lipids for milk synthesis in the dairy cow. A nutrient marker technique similar to that employed by Hachey et al. (1987) was used in which the variation in the naturally occurring ratios of carbon isotopes was used as a marker. Naturally occurring isotopes of carbon include carbon-13 (^{13}C) and carbon-12 (^{12}C). The ratio of ^{13}C to ^{12}C is distinctive for various types of plant matter. Minson et al. (1975) suggested that these differences in carbon isotope ratio could be utilized to study the origin of the composition of cattle tissues. Natural variations in the ratio of stable isotopes in the feed ingested by cattle should be reflected in the composition of the meat and milk of the cattle.

About 99% of all carbon is found as ^{12}C isotope, while about 1% is found as ^{13}C. The ratio of isotopes will vary depending on the plant material (Minson et al., 1975). Plants that fix carbon dioxide by the Calvin (C_3) pathway have lower ^{13}C:^{12}C than plants that utilize the dicarboxylic acid or C_4 pathway. C_3 plant sources include rye grass (*Lolium* spp.), white clover (*Trifolium repens*), and barley meal. C_4 plant sources include paspalum (*Paspalum dilatatum*) hay, maize silage (*Zea mays*), and meal. Minson et al. (1975) hypothesized that if cows that had previously grazed only C_3 plant sources were suddenly changed to a diet consisting of C_4 plant sources, the origin of the carbon (feed or body tissues) could be estimated by measuring the isotope concentrations in milk.

Six cows were grazed on C_3 pasture and C_3 stall diets for an extended period, fasted for 24 h to encourage clearance of C_3 food residues, and then transferred abruptly to a C_4 diet for a period of 8 or 9 days. Wilson et al. (1988) found that a large proportion, from 43 to 54%, of the carbon in milk fat in early lactation originated from body tissues (C_3 label). This range is strikingly close to the percentage found by Hachey et al. (1987) in their study of lactating women. In late lactation, the proportion of carbon in milk fat that originated from body tissues dropped to 19%. Late lactation in cows represents the time period when milk has been established and body weight loss has stabilized. Although the study populations were small, the studies of Hachey et al. (1987)

and Wilson et al. (1988) both suggest that an appreciable portion of fats in milk arises from the mobilization of adipose tissue.

West et al. (1989) studied the mobilization of adipose tissue stores after a meal in the rat. They found that the flux of lipid into and out of adipose tissue stores occurs rapidly, and is a normal function of changes induced by eating. During and immediately after a meal in humans and other omnivores, the organism adjusts its physiology and metabolism to digest, absorb, and utilize ingested nutrients and to store calories that are not immediately required to meet energy needs. In the postabsorptive phase, calories are stored as glycogen in liver and muscle, or as lipid in liver and adipose tissue, until subsequently mobilized to provide needed energy until the next meal.

Adipose tissue has traditionally been viewed as an energy storage depot for long-term caloric needs. The work of West et al. (1989), however, suggests that the dynamics of adipose tissue mobilization appear to be much more responsive to short-term energy demands. Although the study was conducted using adult male rats, the results should be applicable to adipose tissue response to energy demands in general, and may be even more relevant to the additional demands on energy during lactation.

Chappell et al. (1985) studied the influence of maternal diet and weight loss during lactation on the composition of lipids in human milk. Levels of trans fatty acids range from 2 to 18% of total fatty acids in human milk. The level of trans C_{18} (9) fatty acid was found to range from 0.5 to 4.5% of total fatty acids in human milk in this study. This fatty acid has been reported to be the major fatty acid of adipose tissue, regardless of the composition of fatty acids in dietary intake (Field et al., 1985, as reported by Chappell et al., 1985).

Chappell et al. (1985) found that women who had a high weight loss (4–7 kg) in the first 5 weeks of lactation had a significantly higher baseline level of milk trans fatty acids ($p < .001$) compared to women with a low weight loss (0–2 kg) during the same period of lactation (exact figures were not provided). In addition to adipose tissue, trans fatty acids are also present in low concentrations in the diet, but the differences observed in this study indicate the adipose tissue stores were a significant source of the increased trans fatty acids. The milk volumes produced and energy consumed by the mothers in the high and low weight loss groups were not significantly different. The authors indicate that none of the 14 mothers stated that they were actively dieting during this time period. The higher weight loss group may have differences in metabolic rates, daily activity levels (energy demands), and other factors compared to the low weight loss group, which lead them to use adipose tissue for energy, independent of caloric intake.

The increased trans fatty acid content in the milk may indicate that active mobilization of adipose tissue is occurring. As noted by others (Lawrence, 1989; Gallenberg & Vodicnik, 1989), when the mother is calorie deficient, depot fats are mobilized and the characteristics of the milk resemble depot fat. However, the data of Hachey et al. (1987) and Chappell et al. (1985) [and indirectly, the

data of West et al. (1989) and Wilson et al. (1988)] indicate that the mobilization of adipose tissue stores takes place even when the mother has adequate
caloric intake, although a further degree of adipose mobilization may take place
if the energy requirements are not met. With the mobilization of adipose tissue
to supply the lipids for milk, mobilization of lipid-soluble toxicants stored in the
fat is likely to occur as well. These studies did not investigate the occurrence or
partitioning of fat-soluble contaminants in adipose tissue or milk.

The lipid concentration of milk varies within a single nursing period, showing sharp end-feed peaks (Woodward et al., 1989). In a study of 13 lactating
women, Woodward et al. (1989) found that the mean and standard deviation of
grams of fat per liter of milk at the start of a feed was 30 ± 8 and 37 ± 9 g/L for
the first breast and second breast, respectively ($p < .01$). At the end of a feed,
the g fat levels were 50 ± 10 and 47 ± 9 g/L for the first and second breast,
respectively ($p < .01$). The authors hypothesize that the rise in fat content at the
end of a feed is due to the dislodging of the more tightly bound fat at the
terminus of the mammary alveoli as milk is removed from the breast. The fat
content at interruption of a feed was 39 ± 9 and 41 ± 9 g/L, respectively, for
first and second breast. Generally, the most important predictor of fat content is
the length of time since the last feed; the longer the interval, the lower the fat
concentration is. Also, the larger the milk consumption at a previous feed, the
greater is the increase in fat from the beginning to the end of the feed (Lawrence,
1994).

In summary, the fats in breast milk arise from three sources: about 30%
from the current maternal diet, 60% from maternal adipose tissue stores and
10% from de novo synthesis in the mammary gland. The contribution from
adipose tissue takes place despite adequate maternal caloric dietary intake. Adipose tissue may contain chemicals to which the mother has been exposed prior
to and during lactation. These chemicals are mobilized when adipose tissue is
called upon to supply the lipids for breast milk. Women who lose weight rapidly during lactation may be contributing greater amounts of lipophilic contaminants to breast milk.

The lipid concentration of breast milk varies within a single nursing period,
over time, and with subsequent offspring. Although differences in the maternal
diet lead to differences in some fatty acids in breast milk, the composition of
breast milk is remarkably constant. The lipids in infant formula are usually
provided by vegetable oils, and are generally free from chemical contamination.
Water used to reconstitute formula, however, may contain environmental chemicals
if the water is from a contaminated source.

PASSAGE OF CHEMICALS TO BREAST MILK

Several recent reviews have evaluated the factors that influence the transfer of
drugs and environmental chemicals to milk (Gallenberg & Vodicnik, 1989; Atkinson
et al., 1988; Gardner, 1987; Rivera-Calimlim, 1987; Lawrence, 1989). Much of

what is known regarding the passage of chemicals to milk has been generated by studies of drug transfer. The medical community has long recognized the potential impact of maternal medication on the nursing infant via breast milk. Most drugs administered to lactating mothers are excreted to some extent into breast milk. As the analytical capability to detect trace concentrations has improved, increasing numbers of drugs and environmental chemicals have been detected.

Physicians usually advise pregnant or lactating women to avoid the use of therapeutic agents unless the drugs are essential for maternal and/or fetal/infant well-being. This recommendation is generally based on the lack of information demonstrating safety, rather than the existence of data showing harm (Gallenberg & Vodicnik, 1989). With very few exceptions (digitalis, for example), less than 1% of a maternal dose of a therapeutic agent is transferred to the nursing child (Lawrence, 1989; Atkinson et al., 1988). However, the physicochemical characteristics of environmental contaminants are quite different from those of therapeutic agents. Generally, therapeutic drugs are relatively hydrophilic, are readily metabolized and excreted, and do not distribute to adipose tissue to an appreciable extent (Gallenberg & Vodicnik, 1989). If the drug is believed to have potentially adverse effects on the neonate, exposure to the agent can usually be avoided.

Many environmental chemicals, however, are ubiquitous in the environment, are not readily metabolized or excreted, and may accumulate in the maternal body for the 15–45 years prior to a woman's pregnancy. Some may even have been transferred to her by the milk of her own mother. In comparison to therapeutic medications to which the patient is knowingly exposed, the dose, route, and duration of exposure to environmental chemicals are often unknown and may not be recognized as taking place.

The factors that determine the transfer and ultimate concentration of a chemical in breast milk can be divided into three groups: chemical characteristics, maternal characteristics, and maternal exposure magnitude. The most important chemical characteristics are (1) ionization (as measured by the plasma-to-milk pH gradient), (2) lipid solubility (as measured by the octanol:water partition coefficient), (3) molecular weight, and (4) association with blood constituents. The most influential maternal characteristics include (1) milk fat content, (2) lipid genesis (adipose and dietary), (3) maternal parity, (4) maternal adipose tissue mass, and (5) maternal physiology (metabolism and excretion, cardiac perfusion, and pulmonary ventilation). Maternal exposure magnitude, prior to and during lactation, is the major determinant of the concentration of a chemical in breast milk. The mother's age is an important determinant since many persistent chemicals are stored in maternal tissues and build up over time. Adipose tissue contaminant concentrations are an index of the degree to which the mother has been exposed to lipophilic chemicals. Adipose tissue is an important contributor of breast-milk contaminants since it provides about 60% of the fats in breast milk. These factors are discussed in detail next.

Chemical Characteristics

Degree of Ionization   Most chemicals are either weak acids or weak bases, and when they dissolve in a fluid such as blood or milk, they dissociate into two forms: ionized (containing an electrical charge) and nonionized (neutral). Only the nonionized form can readily pass into milk because the lipophilic character of the cell membrane does not allow charged particles to pass through it unassisted (Gardner, 1987). The degree to which a chemical dissociates is determined by the pH of the fluid in which it is dissolved and the dissociation constant (pK_a) of the chemical. Breast milk is more acidic (pH ranges from 6.8 to 7.3) than plasma and interstitial fluid (pH 7.4) (Gardner, 1987; Lawrence, 1989). Chemicals that are weak acids (low pK_a) are ionized to a greater extent in alkaline solution and will have a higher concentration in plasma than in milk. Conversely, weakly alkaline compounds (high pK_a) will have higher levels in milk than in plasma (see Table 5-1).

The relative concentration of a chemical in milk and plasma, referred to as the milk to plasma ratio (M/P ratio), is commonly used to express the relative concentrations of a chemical in milk compared to the concentration in maternal plasma. Most therapeutic drugs have an M/P ratio considerably less than 1; however, this is not the case for most chemicals that are environmental contaminants. Persistent environmental contaminants are chemicals that are found in the

**Table 5-1   Milk to Plasma Ratios for Selected Drugs
and Environmental Contaminants**

Drug	M/P ratio	Reference
Aspirin	0.03–0.08	Lawrence (1989)
Sulfacetamide (low pK_a)	0.08	Lawrence (1989)
Sulfanilamide (high pK_a)	1.00	Lawrence (1989)
Nadolol	4.6	Lawrence (1989)
Morphine	2.5	Atkinson et al. (1988)
Tetrachloroethene	3.2 mean	Sheldon et al. (1985) (range 0.5–6.7)
	3.3	Bagnell and Ellenberger (1977)
PBB	3	Lawrence (1989)
PCBs	3.9	Sheldon et al. (1985)
	4–10	Lawrence (1989)
DDE	12.5	Lawrence (1989)
	6–7	Lawrence (1989)
Dieldrin	6	Lawrence (1989)
Heptachlor epoxide	2–9	Lawrence (1989)
Methylmercury	8.6	Lawrence (1989)
Cadmium	0.54	Sheldon et al. (1985)
	0.12 (nonsmokers)	Radish et al. (1987)
	0.10 (smokers)	Radish et al. (1987)

Note: For many environmental chemicals there is inadequate information on milk and plasma concentrations measured concurrently.

environment by virtue of their manufacture, use, and distribution by people. Contaminants, like DDT, are organic molecules with little ionization potential and a high degree of lipophilicity. These factors combine to result in M/P ratios that approach and exceed 1.

The M/P ratios are helpful to assess whether high concentrations of a therapeutic drug are expected in breast milk when a lactating woman is medicated. Often, the maternal plasma concentration of a therapeutic drug can be measured or is known based on previous therapeutic experiences, and the milk concentration can be estimated using M/P ratios based on human or animal data, or predicted by pharmacokinetic models. The M/P ratios for environmental contaminants can be used as an index of the likelihood of detecting the chemical in breast milk subsequent to exposure.

Lipid Solubility   The lipid solubility of a chemical is an important determinant of the concentration expected in breast milk. Since human milk contains a 4% lipid phase (Lawrence, 1989), lipid-soluble chemicals may partition into milk fat and result in higher concentrations in milk than in plasma, which contains about 0.65% circulating blood lipids (Task Group on Reference Man, 1974). Whereas ionized chemicals are prevented from traversing the lipid-containing cell membranes, lipid-soluble chemicals rapidly cross this barrier. In addition, since breast milk is periodically removed from the mother by the nursing infant, the concentration gradient favors additional transfer of the lipid-soluble chemical present in the circulating blood lipids and adipose tissue stores. Lipid solubility is often assessed by its octanol–water partition coefficient (K_{ow}). The K_{ow} is the ratio of the solubility of a chemical in octanol (lipid soluble) to the solubility of a chemical in water. The higher the K_{ow}, the more the chemical will partition to lipids relative to aqueous solution.

Molecular Weight   Molecular weight is a factor in the transport of water-soluble chemicals across cell membranes. Water-soluble chemicals generally traverse the cell membrane in water-filled pores that restrict high molecular weight molecules from transport, by virtue of their size. It is believed that heparin and insulin, for example, are not found in human milk due to this apparent size restriction (Lawrence, 1989). This does not apply to lipid-soluble chemicals. High-molecular-weight lipid-soluble molecules such as PCBs and DDT readily cross the cell in the lipid phase of the membrane. The lipophilic character of these chemicals enables them to cross the membrane of the alveolar epithelial cells, which are comprised of lipoprotein, glycolipid, phospholipid, and free lipids.

Association with Blood Constituents   In addition to simple diffusion of molecules (which is governed by the concentration gradient between milk and plasma), chemicals may also be passed into milk by carrier-mediated diffusion and by active transport. Pinocytosis and reverse pinocytosis are involved in the carrier-mediated diffusion of some very large biologically active molecules and proteins (Gardner, 1987). Active transport requires energy and permits substances to move from lower concentrations to higher concentrations. Substances that are actively transported include glucose, amino acids, calcium, and magnesium (Gardner, 1987).

In addition to lipophilic transport, chemicals may be bound to proteins in blood and milk or may remain free in solution. The amount of a chemical bound to protein in milk is less than the amount bound to protein in plasma, consistent with the lower protein content of milk (8 to 11 g/L) compared to plasma (74.6 g/L). A high degree of protein binding in plasma effectively restricts the movement of the chemical to the milk. However, carrier-mediated diffusion and active transport mechanisms may allow higher concentrations of certain substances in the milk than the plasma without regard to the concentration gradient.

Maternal Factors

The degree to which contaminants are present in breast milk is associated with the way in which the contaminants are absorbed, distributed, metabolized, stored, and excreted by the lactating woman. These maternal factors include milk fat content, genesis of milk fats, maternal parity, adipose tissue mass, and maternal physiology.

Milk Fat Content Most chemicals found in breast milk in relatively high concentrations are lipid soluble and poorly metabolized. The concentration of these chemicals is primarily associated with the amount of lipid in breast milk. The higher the lipid content, the higher is the concentration of lipid-soluble chemicals expected in breast milk. Fats are the most variable constituents in human milk, varying in concentration over a feeding, over a day's time, and over the course of a complete lactation period (Lawrence, 1989).

The fat content of mature human milk increases from a lower concentration at the beginning of a single nursing session to a higher concentration at the end. This is thought to arise from the dislodging of tightly bound lipid globules from the mammary ducts (alveoli) as milk is removed from the breast. Increases from beginning to end of a single nursing session have been cited as ranging from 6 to 79% (Jensen, 1989). Results found by different investigators are often difficult to compare reliably due to the variety of analytical techniques employed to measure fat content.

A typical median fat content of 3.9% at the beginning of a nursing session was found to increase to 6.1% at the end in extensive studies conducted by Macey et al. (1931, as reported by Jensen, 1989). Newer studies designed to measure milk fat variation are not available. The fat content of human milk is usually cited as 2.5–4.5%, but is known to range widely. Mature breast milk contains an average fat content of 3.9% according to a recent review by Jensen (1989). Four percent milk fat is assumed for the calculations in this document.

The stage of lactation also has an influence on the lipid content. Colostrum is the "early milk" established from birth until about 5 days postpartum and has an average fat content of 2.9%. Colostrum is followed by transitional milk from days 6 to 10, with an average fat content of 3.6%. As the mother establishes her

milk supply and the milk becomes mature, the fat content levels off and there-
after stabilizes with small declines or increases as lactation progresses to what
appears to be individually predisposed levels (Jensen, 1989).

Diurnal variations in milk fat content have also been reported, but appear to
occur at different times of day depending on the population studied. In a group
of Western women, the peak fat content occurred at about 6 p.m., while in a
group of Third World women the peak occurred in the early morning (Jackson
et al., 1988).

Lipid Genesis   The genesis of the fats contributing to the lipids in human
milk will also influence the degree to which contaminants are present in human
milk, as previously discussed. If the maternal adipose tissue stores are contaminated with environmental chemicals, these contaminants will be present in the
milk since the adipose fats contribute about 60% of the milk fat lipids. Similarly, the contaminants in the foods ingested by the mother will be present in the
milk since the dietary fats contribute about 30% of the milk fat lipids.

Maternal Parity   The number of children previously nursed by a mother
will affect the chemical levels in her milk for her next offspring. Breast milk
may be thought of as an excretory route for lipid-soluble, poorly metabolized
chemicals. Because little is lost through urine and feces, the removal of this type
of chemical in milk is often the most effective route of maternal excretion.
Subsequent offspring will be nursed with breast milk containing lower concentrations of persistent contaminants unless extraordinary exposure has occurred
between pregnancies (Lawrence, 1989). Parity also has a major influence on fat
content. The milk from primiparous mothers has more fat than milk from mul-
tiparous mothers (Prentice et al., 1981; Macey et al., 1931).

Maternal Adipose Tissue Mass   Since adipose tissue supplies about 60%
of the lipids in breast milk, the amount of adipose tissue and the contaminant
load in the adipose tissue will affect the contaminant concentration in breast
milk. The body fat mass of lactating women has been measured by a number of
investigators. Wong et al. (1989) studied body composition of 10 lactating women
by anthropometry and by deuterium dilution methods. Both methods estimated
about 295 g of body fat/kg body weight (30%), or 18.2 kg fat in an average 62-
kg lactating woman. Obese women will have a significantly greater body fat
mass (up to 50%) than women who are at or near their optimal body weight.
This extra fat provides a greater reservoir into which fat-soluble contaminants
are stored, leading to a "dilutant" effect assuming similar exposure to contaminants. However, if the heavier woman ingests greater amounts of food containing contaminants, then the concentrations of a contaminant in fat are likely to be
similar to the levels found in the general population. There are no studies available that examine contaminant concentrations in the fat of obese women compared to women in the general population.

Maternal Physiology   Pregnancy and lactation are physiological states during which the tissue distribution of lipophilic chemicals may be altered due to
modification of the lipid-to-protein ratios of plasma (Gallenberg & Vodicnik,

1989). Pregnancy, in both rodents and humans, results in a hypertriglyceridemic state during the third trimester, in preparation for milk fat synthesis. The very-low-density lipoprotein (VLDL) pool is enhanced with advancing pregnancy in humans and rodents, and this increase appears to be related to both its estrogen-stimulated secretion by the liver and its reduced clearance due to an inhibition of adipose tissue lipoprotein lipase activity. VLDL is a major substrate for lipo-protein lipase, which is decreased in adipose tissue but elevated in the mammary gland during late pregnancy and lactation to supply free fatty acids for milk TG synthesis. Total triglyceride levels during weeks 34–38 of human pregnancy have been shown to be 3 times higher than those in nonpregnant women (Montes et al., 1984). Total cholesterol was also elevated by 50%. In a study comparing lipoprotein lipids and apoproteins for lactating (n = 56) and nonlactating (n = 16) women 6 weeks postpartum, Knopp et al. (1985) found significant differences in several blood lipids. Notably, in lactating women, levels of total and high-density lipoprotein cholesterol were significantly higher, while levels of total and very-low-density lipoprotein triglycerides were signifi-cantly lower than in nonlactating postpartum women. Also, in lactating women, levels of high-density lipoprotein apoprotein A-I and A-II were significantly higher, while levels of high-density lipoprotein phospholipids and very-low-density lipoprotein phospholipids were significantly lower than in nonlactating postpartum women.

Lipophilic chemicals that are released from adipose tissue to the blood dur-ing lipid mobilization (or as a result of ingestion or inhalation from environ-mental sources) become immediately associated with circulating VLDL in the pregnant or lactating woman, or may also be delivered to the liver where they are packaged into newly synthesized VLDL. These physiological modifications may help explain the apparently higher retention of inhaled volatile organic compounds by the lactating woman compared to the nonlactating woman in studies conducted by Wallace et al. (1989) and Sheldon et al. (1985).

The degree to which the chemical is metabolized by the mother will also influence whether the chemical is ultimately found in breast milk. Generally, chemicals are metabolized to more water-soluble metabolites and are then ex-creted via urine or feces. Unmetabolized lipid-soluble chemicals remain in the maternal blood circulation and eventually become deposited in adipose tissue. Their presence in the bloodstream is usually associated with the VLDLs from dietary sources and adipose tissue, which become incorporated in breast milk during lactation.

Maternal Exposure

The magnitude and duration of maternal exposure influence the amount of the chemical in breast milk. Once the chemical has been absorbed and distributed in the maternal bloodstream, it is available for distribution to breast milk, usually via the circulating plasma lipids (for fat-soluble contaminants) or via the circulating plasma proteins (for protein-associated contaminants). The equilibrium between chemical concentrations in plasma and milk may be disturbed due to the variable chemical plasma levels (which reflect the current exposure as well as the contribution from adipose tissue mobilization) and the removal of the contaminated milk fats by the nursing infant.

As the infant nurses and removes milk (and milk fats) from the mother, the equilibrium shifts because the chemical concentration in the mammary fluids is reduced. This will enable more contaminants to be drawn from the blood plasma and adipose tissue stores until equilibrium is reestablished. If one assumes a constant maternal exposure (via inhalation or ingestion), the contaminant will continue to move into breast milk as the milk is periodically removed from the breast by the nursing infant. The greater the maternal exposure, the higher the concentration in blood plasma, and consequently the higher the expected concentration in breast milk.

The mother's age influences the degree to which chemicals are present in her milk. Generally, the concentrations of persistent, lipophilic chemicals increase with increasing maternal age. This is a function of the longer exposure duration the older mother has had over her lifetime. The trend of increasing chemical concentration with age implies that low-level dietary and other exposures continue to be a source of maternal contamination. If the sources of maternal exposure to contaminants were curtailed or reduced, this phenomenon would probably not occur. Conversely, the introduction of newly synthesized chemical products to the marketplace or home can be a source of unrecognized contamination if surveys to monitor their presence in environmental or biological samples have not yet been conducted.

One index of maternal exposure is the concentration of chemicals in adipose tissue. Analysis of body fat for environmental contaminants has revealed the presence of many persistent chemicals (PCBs, DDT, etc.) as well as the presence of volatile and semivolatile organic compounds. Analysis of adipose tissue in the United States has been conducted by the National Human Adipose Tissue Survey (NHATS), a monitoring program of the Office of Toxic Substances within the U.S. Environmental Protection Agency (EPA). The objective of the NHATS program is to detect and quantify the prevalences of toxic compounds in adipose tissue in the general population.

Adipose tissue specimens were collected during surgical procedures or as part of postmortem examination according to a planned statistical design (Lucas et al., 1981). The survey design ensures that specified geographical regions and demographic categories were appropriately represented to allow estimates of baseline levels, time trends, and comparisons across subpopulations. The cooperating physicians and pathologists acquired at least 5 g of subcutaneous, perirenal, or mesenteric high-lipid adipose tissue, taking precautions to avoid contamination that might result from direct exposure to chemicals such as solvents, paraffin, disinfectants, preservatives, or plastics (U.S. EPA, 1986a). Details of the procedures for collection, compositing, and analysis of samples can be found in a series of U.S. EPA documents (U.S. EPA, 1986b, 1986c, 1986d, 1986e, 1987a, 1987b, 1989a).

Nine geographical census tract areas were evaluated: New England, Middle Atlantic, East North Central, West North Central, South Atlantic, East South Central, West South Central, Mountain, and Pacific. New York State is represented in the Middle Atlantic census tract area. The samples were sorted by age (0–14 years, 15–44 years, and greater than 45 years), but not by sex. The adipose tissue sample composites were collected randomly from individuals having no known occupational exposures, and should be representative of general population exposure. The NHATS program analyzed 46 composite adipose tissue samples, each comprised of about 20 individual specimens, for the presence of volatile and semivolatile organic compounds at concentrations ranging from 0.001 to 2 µg/g (ppm).

The most prevalent volatile organic compounds and their frequencies of observation were styrene, 1,4-dichlorobenzene, xylene, and ethylphenol (all at 100%); benzene, chlorobenzene, and ethylbenzene (all at 96%); toluene (91%); chloroform (76%); 1,2-dichlorobenzene (63%); and tetrachloroethene (61%) (see Table 5-2). The most prevalent semivolatile organic chemicals and their frequencies of observation were DDE (93%), β-BHC (87%), PCBs (83%), hexachlorobenzene (76%), butylbenzyl phthalate (69%), and heptachlor epoxide (67%) (see Table 5-3). The frequencies of detection and the concentrations at which certain contaminants were identified in NHATS indicate the widespread occurrence and appreciable concentrations of some chemicals in the general population.

Table 5-2 Presence of Volatile Organic Compounds in Composite Adipose Tissue Samples

Compound	Frequency of observation (%)	Wet tissue concentration (ng/g)[a]
Chloroform	76	ND[b] (2)–580
1,1,1-Trichloroethane	48	ND (17)–830
Benzene	96	ND (4)–97
Tetrachloroethylene	61	ND (3)–94
Toluene	91	ND (1)–250
Chlorobenzene	96	ND (1)–9
Ethylbenzene	96	ND (2)–280
Styrene	100	8–350
1,2-Dichlorobenzene	63	ND (0.1)–2
1,4-Dichlorobenzene	100	12–500
Xylenes	100	18–1400
Ethylphenol	100	0.4–400

Note: Data from the National Human Adipose Tissue Survey (NHATS) FY 1982. Data presented for VOCs detected at a frequency of 30% or greater (U.S. EPA, 1986a, 1986b, 1986c). All census regions.
[a]ng/g = ppb.
[b]ND, not detected. Value in parentheses is the estimated limit of detection.

Table 5-3 Presence of Semivolatile Organic Compounds in Composite Adipose Tissue Samples

Compound	Frequency of observation (%)	Wet tissue concentration (ng/g)[a]
Naphthalene	40	ND[b] (9)–63
Diethyl phthalate	42	ND (10)–970
Hexachlorobenzene	76	ND (12)–1300
β-BHC	87	ND (19)–570
Di-n-butyl phthalate	44	ND (10)–1700
Di-n-octyl phthalate	31	ND (9)–850
Heptachlor epoxide	67	ND (10)–310
trans-Nonachlor	53	ND (18)–520
Dieldrin	31	ND (44)–4100
p,p'-DDE	93	ND (9)–6800
Triphenyl phosphate	36	ND (18)–850
p,p'-DDT	55	ND (9)–540
Butylbenzyl phthalate	69	ND (9)–1700
Total PCBs	83	ND (15)–1700

Note: Data from the National Human Adipose Tissue Survey (NHATS) FY 1982. Data presented for semivolatile compounds detected at a frequency of 30% or greater (U.S. EPA, 1986a, 1986b, 1986c). All census regions.
[a]ng/g = ppb.
[b]ND, not detected. Value in parentheses is the estimated limit of detection.

INGESTION EXPOSURE

The presence of residues of pesticides and other organic contaminants in the domestic and imported food supply has been recognized for many years. The U.S. Food and Drug Administration (U.S. FDA) monitors these contaminants in the nation's food supply (Gartrell et al., 1985, 1986). The amount of exposure depends on the contaminant concentration as well as the amount of the food that is ingested by an individual. Differences in intake of various food types by age group (infants, toddlers, teenagers, and adults) result in differences in exposure to contaminant residues. In addition, individual preferences for foods and exposure to locally contaminated foods allow a wide range of individual contaminant intakes.

The lactating woman has been exposed to a variety of contaminants via the foods she has ingested over her lifetime. Fat-soluble, poorly metabolized contaminants such as PCBs, DDT and metabolites, and others not readily excreted remain in adipose tissue in dynamic equilibrium with blood lipids. The adipose tissue concentrations represent the stored contaminant levels that result from current and past exposure via ingestion, inhalation, and dermal routes. The physiology of gastrointestinal lipid digestion and of absorption of lipophilic xenobiotic substances is poorly understood, although knowledge has increased greatly during the last two decades.

There are many fat-soluble molecules that accompany dietary fat during digestion. Some of these molecules are natural substances, such as carotenoids and other hydrocarbons of plants (Borgstrom & Patton, 1991). Most lipophilic xenobiotics and natural products occur as trace contaminants of food or inhaled air. When the inhaled materials contain particulates (e.g., diesel particles), they are cleared from the nasal and bronchial epithelium and are deposited in the upper gastrointestinal tract by mucociliary action (Sun et al., 1984). Fat-soluble molecules in food are dissolved primarily in the fat fraction of the food, particularly in animal products but also in skins and membranes of fruits and vegetables. Most foods are contaminated to some extent by polycyclic aromatic hydrocarbons (PAHs), PCBs, and pesticide residues. Some fish and other fatty animal products may contain significantly elevated levels of PCBs and other contaminants (Zabik et al., 1979; New York State Department of Health, 1991).

In the stomach, fat-soluble molecules associated with plant skins, other natural substances, and xenobiotic contaminants, as well as lipophilic particles from the lungs, are mixed together as the macroscopic structure of food is broken down into microscopic particles during the formation of chyme. An important function of gastric fat digestion appears to be to prepare fat for more efficient and complete digestion by pancreatic lipases in the small intestine. Very little is known about fat-soluble molecule absorption in the stomach.

Although poorly understood, most fat-soluble molecules are extracted from gastric content by dietary fat and enter the small intestine predissolved in triglyceride droplets (Borgstrom & Patton, 1991). Once the dietary fat and the associated fat-soluble molecules enter the intestine, they are codispersed during fat digestion. The dispersion by bile salts is a complex dynamic process that produces a solution in which monomers of nonpolar lipid, digested fat (fatty acids and monoglycerides), and bile salts are in rapidly changing equilibrium with micellar aggregates. During this process, some of the nonpolar lipid molecules may not be able to achieve complete solubility in the micellar solution and will precipitate as insoluble crystals or aggregates to be excreted in the feces. Once dispersed into the micelles, relatively high concentrations of nonpolar lipids and lipophilic molecules can diffuse up the intestinal brush border membrane, where they are absorbed into the bloodstream (Borgstrom & Patton, 1991).

INHALATION EXPOSURE

The physicochemical characteristics of an inhaled agent will influence its deposition and retention in the respiratory tract, translocation within the respiratory system, distribution to blood and other tissues, and ultimately, contribution to breast milk. The deposition and rate of uptake are determined by the reactivity and solubility characteristics of the inhaled material in conjunction with physiologic parameters such as pulmonary ventilation, cardiac output (perfusion), metabolic pathways, tissue volumes, and excretion capacities (Jarabek et al., 1990).

Each minute, an average adult at rest inhales 4–8 L air, which is passed over the lung surface of approximately 70–90 m^2 (Witschi & Brain, 1985; Grunder & Moffitt, 1988). The inhaled air is brought into contact with lung capillaries, where a rapid exchange of water and solutes with interstitial fluid takes place.

The lung capillaries consist of a single layer of epithelial cells through which gases and fluids may be exchanged. The radius of a lung capillary is about 5×10^{-4} cm. The average blood flow in the capillaries is approximately 1 mm/s, although this can vary from zero to several millimeters per second in the same vessel within a brief period of time depending on demands placed on the circulatory system (Berne & Levy, 1981). The heart pumps about 5 L blood per minute in a person at rest; thus the entire blood volume is circulated each 60 s.

Transcapillary exchange of an inhaled gas may take place by diffusion, filtration, and/or transport by endothelial vessels (pinocytosis). The greatest number of molecules traverse the capillary endothelium by diffusion, governed by Fick's law (Berne & Levy, 1981). The concentration gradient between the gas in the capillaries and the tissue cells (blood) will determine the direction of movement of the gas. For nonreactive, lipid-soluble gases, the major factor driving the uptake of the gas is the removal of the gas from the alveolar air by the capillary blood. Until equilibrium is reached, the gas will be drawn from the alveolar air to the bloodstream via the lung capillaries.

The uptake of alveolar gases depends on the air-to-blood partitioning coefficient (chemically dependent), ventilation rate (physiologically dependent), perfusion rate (physiologically dependent), and air and blood concentrations (chemically and physiologically dependent). Therefore, the relationship between inhaled gas concentration in the ambient air to the concentration of the gas in the blood (and later in breast milk) is dependent on both the physicochemical characteristics of the inhaled gas and the physiological state of the person inhaling the gas.

Once the contaminant has been transported across the alveolar epithelium, it enters the systemic circulation. The greater the amounts reaching the systemic circulation, the greater is the likelihood for adverse effects in other systems. The extent to which systemic absorption occurs and the time to reach steady-state blood levels are influenced by (U.S. EPA, 1990):

1 Ventilation rates and airway mechanics
2 Blood transit time in capillary beds (perfusion limited)
3 Metabolic conversion in the respiratory tract and other organs (if applicable)
4 Alveolar surface area
5 Thickness of the air–blood barrier
6 The blood:air and blood:tissue partition coefficients

Inhaled volatile organic chemicals, after absorption in the pulmonary system, are associated with the circulating blood lipids (especially VLDL). When the blood containing these contaminants reaches the lung, some of the volatile organic contaminants may be exhaled, depending on the concentration of the chemical in the newly inspired air. As the blood circulates in the body, some of the contaminant may be metabolized and excreted in the urine or feces. In addition, an equilibrium between blood lipids and adipose-tissue lipids is established. If inhalation exposure were to stop or decrease, some of the adipose-tissue-stored contaminants would reenter the systemic circulation from the adipose tissue, and would be exhaled when the blood–air exchange of gases takes place in the lung. Therefore, the blood lipids maintain equilibrium and act as a conduit between the inhaled air and the adipose tissue. The blood lipids also maintain equilibrium between ingested soluble contaminants and adipose tissue.

Volatile organic chemicals are often present in modern environments due to their use in residential, commercial, and industrial settings. Indoor air and ambient air often have detectable concentrations of these chemicals; certain occupational exposures to selected VOCs can be substantial. Recognition of indoor air as an important exposure pathway has occurred only recently. In studies of contaminant concentrations in apartments near dry cleaners, highly elevated tetrachloroethene concentrations have been determined (Schreiber et al., 1993). Inhalation of contaminated air in these apartments is estimated to result in significant exposure of people residing there, overriding the contribution from contaminated foods and breast milk (discussed later).

Recent unpublished data (Wallace et al., 1989) on the inhalation and exhalation of volatile solvents and their concentration in breast milk suggest that a greater proportion of inhaled solvents is retained by the lactating woman compared to the nonlactating woman, based on comparison of inhaled and exhaled breath concentrations. This difference may be due to greater adsorption of inhaled toxicants by circulating plasma lipids or due to the larger blood volume in the lactating woman, or a combination of these factors. Further examination of this phenomenon is warranted, and may lead to improved ability to predict the presence of substances in breast milk as a result of inhalation of airborne contaminants by the nursing mother. Many of these volatile and semivolatile organic contaminants have also been identified in adipose tissue samples (U.S. EPA, 1986a, 1986b, 1986c, 1986d).

TETRACHLOROETHENE: CASE STUDY

Tetrachloroethene is a volatile, colorless, nonflammable organic solvent widely used in the dry cleaning industry as well as in other applications such as metal degreasing, where a nonpolar solvent is required. Tetrachloroethene (CAS no. 127-18-4), also known as tetrachloroethylene and perchloroethylene (PCE), is relatively insoluble in water (150 mg/L at 25°C) and is miscible with most organic solvents and oils. Its vapor pressure is 14 mm Hg at 20° C.

As is the case with most lipid-soluble volatile organic compounds, after inhalation PCE will attain a concentration in the blood stream. If the PCE concentration in air is greater than the PCE concentration in the blood at the beginning of the exposure period, when the inhaled air carrying PCE is brought into contact with blood in the alveolar capillaries, PCE will be absorbed into the blood by passive diffusion. The amount of PCE absorbed is a percentage of the amount of PCE inhaled. As all of the PCE is not absorbed and little is metabolized (less than 5%), the remainder is exhaled unchanged or remains in adipose tissue stores. By knowing the concentration of PCE in the inhaled (ambient) air as well as the concentration of PCE in exhaled breath, one can estimate the amount retained by the body with each breath.

As exposure continues, the inhaled air containing PCE will continue to diffuse PCE to the blood until an equilibrium is established at the air–blood interface. The equilibrium concentration is determined by the physicochemical characteristics of the VOC and the physiological parameters associated with the particular exposure. The most important physicochemical characteristic is the air-to-blood partitioning coefficient, which is itself governed by the solubility of the material in air and in blood (see Table 5-4). A partition coefficient is the ratio of concentrations achieved between two different media at equilibrium (Gargas et al., 1989). Important physiological parameters include cardiac per–fusion rate, inhalation rate, and energy expenditure of the person exposed. When equilibrium has been reached, there will be an equal average number of molecules transferred across the alveolar capillary epithelium in both directions. That is, PCE will be given up by the air to the blood in the same amount as will be given from the blood back to the alveolar air (washout) whence it is exhaled. After inhalation exposure has ended, the blood will continue to diffuse PCE across the alveolar epithelium to the lungs. Rapid initial elimination is followed

Table 5-4   Human Partition Coefficients for PCE

Partition coefficients	Reference
Blood:air partition coefficient, $\lambda(b/a)$, liters air/liters blood	
10.3	Travis et al. (1989)
15	Monster et al. (1979)
13.1	Sato and Nakajima (1979)
9.1	Morgan et al. (1970)
18.9	Fiserova-Bergerova (1983)
10.3	Bogen and McKone (1988)
$\lambda(b/a)$ ave $= 12.8$, $\lambda(a/b) = 0.078$	
Fat:air partition coefficient, $\lambda(f/a)$, liters air/liters fat	
1638	Travis et al. (1989)
1917	Sato and Nakajima (1979)
$\lambda(f/a)$ ave $= 1777$	
Fat:blood partition coefficient, $\lambda(f/b)$, liters blood/liters fat	
159	Bogen and McKone (1988)

by a slower phase with a half-life of 34–72 h. The storage of PCE in adipose tissue accounts for its slow excretion (Illing et al., 1987), and also accounts, in part, for its presence in breast milk.

The only report available in the literature on PCE effects via exposure to breast milk is a clinical case report by Bagnell and Ellenberger (1977). They report a case of a 6-week-old breast-fed infant with obstructive jaundice (conjugated hyperbilirubinemia) and hepatomegaly (see page 121).

PCE is excreted in the milk of exposed animals and in human breast milk (Wanner et al., 1982, as reported by Tabacova, 1986; Bagnell & Ellenberger, 1977), and may be a significant route of PCE exposure for nursing infants. In cows, PCE was absorbed following administration in the feed and identified as present in blood and milk, although quantitative estimates were not provided (Wanner et al., 1982, as reported by Illing et al., 1987).

Maternal Inhalation Exposure

Since PCE is used extensively in occupational settings, exposure of certain workers to PCE can be considerable. At greatest risk of exposure are dry cleaning workers, especially those involved in transfer of dry cleaned garments during the cleaning process. PCE is also used in degreasing metals, as a heat-exchange fluid, as a solvent for silicones, and as an insulating fluid and cooling gas in electrical transformers (OSHA, 1989a). No information is available regarding the number or percent of dry cleaning facility workers who are women, nor the number who may be pregnant or lactating. Historically, however, the dry cleaning industry has a large proportion of female workers, often members of family-operated businesses.

The dry cleaning process involves three stages: cleaning, extracting, and drying (Materna, 1985). During the cleaning cycle, fabrics are agitated in the solvent to dissolve and remove soil. Extraction occurs as the fabrics are spun rapidly to remove excess solvent. Drying takes place by tumbling the fabrics in a heated air stream to evaporate any remaining solvent. Solvent vapors are recovered by condensation and purified by filtration for reuse.

There are two basic types of dry cleaning equipment: transfer or dry-to-dry. With transfer-type equipment, the garments are cleaned in a washer/extractor unit and are manually transferred to a separate dryer. The dry-to-dry process is essentially a closed system. The operator loads dry clothing into the machine and the clothing is not removed until the entire cycle is completed. When the fabrics are removed from the machine, they are dry and essentially free of solvent. The more sophisticated dry-to-dry machines have refrigerated condensers and/or carbon absorbers to recapture tetrachloroethene. Clearly, the transfer-type equipment allows more opportunity for exposure of the operator. Estimates of worker exposure to PCE are given in Table 5-5.

The general population is also exposed to PCE. People living in rural locations are exposed to low levels of PCE ranging from 0.008 to 0.5 $\mu g/m^3$, whereas

Table 5-5   Estimates of Occupational Exposure to Tetrachloroethene in Dry Cleaning Establishments

Job or sample description	Number of shops/samples	Air PCE concentrations, mg/m³ range	Arithmetic mean	Geometric mean
Machine operator, TWA	44/45	27.6–1026.6	213.6	151.6
Presser, TWA	35/52	0.69–254.9	39.3	22.7
Seamstress, TWA	12/12	4.1–199.8	45.5	20.7
Counter area, TWA	31/31	2.1–179.1	40.7	21.4
Machine operator, 5 min peak	39/134	22.7–2521.7	523.6	303.2
Machine operator, 15 min peak	39/49	6.9–1853.4	379.0	227.4
Area of clothing transfer, TWA[a]	17/17	196.4–2085.6	596.7	
Area of clothing transfer, 5 min peak[a]	17/17	77.9–3677.9	936.4	

Note: Data from Ludwig et al. (1983) unless otherwise noted. TWA, time-weighted average. NA, not available.
[a]Data from Materna (1985).

people living in U.S. urban areas are exposed to levels ranging from 0.2 to 52 μg/m³ (WHO, 1984). The U.S. EPA maintains a VOC (volatile organic chemicals) national ambient database, which compiles the relevant data on outdoor and indoor levels of volatile organic compounds (Shah & Singh, 1988). The database includes concentrations measured at rural, suburban, and urban locations as well as source-dominated sites. Based on 3226 data points, the arithmetic mean and median ambient (outdoor) PCE concentrations are 5.9 and 2.4 μg/m³, respectively. Based on 2195 data points, the arithmetic mean and median indoor PCE concentrations are 21.1 and 5.1 μg/m³, respectively. The mean concentrations reported may be skewed high due to the inclusion of a few high values. Both the average and median PCE concentrations are greater indoors than the concentrations found outdoors. This is consistent with the existence of strong indoor sources of PCE, a majority of which is thought to be associated with volatilization of PCE from freshly dry-cleaned clothes. PCE residues have been identified in clothing that has been dry-cleaned, and off-gassing from clothing can impact indoor air quality, resulting in measurable PCE concentrations (Thomas et al., 1989; Kawauchi & Nishiyama, 1989; Tichenor et al., 1988).

People living near dry-cleaning establishments can be exposed to considerably elevated PCE concentrations. An investigation by Verberk and Scheffers (1979) found that the breath of residents living above 12 dry cleaning shops in the Netherlands contained a mean concentration of 5000 μg/m³, while the breath of residents living adjacent to the shops contained 1000 μg/m³.

The New York State Department of Health conducted an investigation to determine if tetrachloroethene levels in the indoor air of residences located in the same building as dry cleaning facilities were higher than levels in residences not near a dry cleaner. Data were also collected to evaluate what cleaning equipment or other factors might be contributing to air contamination in the dwellings (Schreiber et al., 1993).

Six resident-occupied apartments above dry cleaning facilities were evaluated. Six additional apartments that had similar building and neighborhood characteristics without a nearby source of tetrachloroethene were selected as controls. At each location, both indoors and outdoors, two consecutive 12-h air samples were collected: the first from 7 a.m. to 7 p.m. (referred to as the AM sample) and the second from 7 p.m. to 7 a.m. (referred to as the PM sample). All samples for a study residence and its control were collected concurrently. A wide variety of conditions within the dry cleaning establishments were found. Three of the dry cleaners used transfer machines, two used newer dry-to-dry machines, and one used a very old dry-to-dry machine in poor operating condition. This dry-to-dry machine was considered separately from the other dry-to-dry machines.

Clearly elevated levels of tetrachloroethene were found in the indoor air of the apartments located above each of the dry cleaners in the AM samples (range 300–55,000 $\mu g/m^3$) compared to the control residences (range <6.7–103 $\mu g/m^3$). Similar results were found in the PM samples, where concentrations of tetrachloroethene in the study residences (range 100–36,500 $\mu g/m^3$) also greatly exceeded the concentrations in the control residences (<6.7–77 $\mu g/m^3$). Although air concentrations in the apartments were usually less at night than during the day, the study residences always had higher concentrations of tetrachloroethene than the control residences. The tetrachloroethene concentrations in outdoor air near the dry cleaners were also significantly elevated compared to control locations away from the dry cleaners, and these levels were less than the indoor levels.

The type of dry cleaning machine was significantly associated with the concentration of tetrachloroethene found in the apartment above, even though only six residences were evaluated. The tetrachloroethene levels in the apartments above dry cleaners using transfer machines were significantly elevated (AM range 1730–17,000 $\mu g/m^3$ and PM range 1350–14,000 $\mu g/m^3$) compared to those using newer dry-to-dry machines (AM range 300–440 $\mu g/m^3$ and PM range 100–160 $\mu g/m^3$). The apartments above the old dry-to-dry unit had the highest concentrations of all (AM 55,000 and PM 36,500 $\mu g/m^3$).

Food and Water

In addition to airborne exposures, the general population may also be exposed to PCE via ingestion of contaminated food and drinking water. Fawell and Hunt (1988) cite a European report that states that PCE is used to extract oil and fat from meat, which may then become contaminated via this process. Levels of PCE of 1 µg/kg have been detected in beef, up to 2 µg/kg in fruit and vegetables, 3 µg/kg in tea and coffee, and in fats from 7 µg/kg in olive oil to 13 µg/kg in butter (McConnell et al., 1975, as reported by Fawell & Hunt, 1988).

Elevated levels of PCE in margarine and butter samples from food stores located near dry cleaning facilities have been reported by FDA researchers. Entz and Diachenko (1988) identified four samples of margarine with elevated PCE levels among a group of samples collected for routine monitoring. The samples were found to come from a store located immediately next to a dry cleaning establishment. Follow-up investigation (Miller & Uhler, 1988) was conducted to determine the levels of PCE (and other VOCs) in fatty foods purchased from stores located both near and distant from dry cleaners. Butter was chosen as the food to monitor due to its uniform very high fat content and loose packaging (most often in waxed or foil wrapper in a paper container), which allows indirect contact with ambient air.

Forty-six packages of butter were purchased from food stores (14 retail outlets) located next to dry cleaners (3 stores), one to two stores away from dry cleaners (6 stores), or not near a dry cleaner (5 stores). As hypothesized, butter samples obtained from stores located near dry cleaning establishments contained elevated levels of PCE (range of 100 to greater than 1000 µg/kg, exact figures not presented). Butter samples from stores near, but not adjacent to, dry cleaning establishments also had elevated PCE levels (range of less than 50 to over 1000 µg/kg). This category included both large and small retail food establishments, frequently in small suburban shopping centers located in the same overall mall structure.

PCE levels in butter samples from retail stores away from dry cleaners were considerably lower (range less than 50 to 111 µg/kg). These data exemplify the vulnerability of fatty foods to PCE contamination from fugitive emissions of dry cleaning establishments.

The residents who live near an active dry cleaning establishment are exposed to PCE via the inhalation of contaminated air in their dwellings. In addition, the airborne PCE is likely to be absorbed by fatty foods that come in contact with the air in the residence. Of particular importance are butters, cooking oils, and margarine, which are composed primarily of fat. The FDA investigations found very high levels of PCE in butter samples collected from retail stores located near dry cleaning establishments. PCE concentrations of greater than 1000 µg/kg (ppb) were identified from butter samples in stores adjacent to the dry cleaning establishments. Importantly, the butter samples came from packaged butter as one would find on store shelves, and not butter on a counter left open to the air. This suggests that the PCE is able to penetrate the wrapper and contaminate the butter even though it is enclosed in the normal retail packaging.

A lactating woman consumes approximately 92.4 g fats/day, assuming a total dietary intake of 2500 kcal/day (NRC, 1989) of which 30% is provided by fats (9 kcal/g). If all the fats a lactating woman consumes are contaminated to the same extent as the butter in the preceding example (a high estimate), then she will consume an additional 92.5 µg of PCE/day. In addition to the inhaled PCE and the PCE contained in the fats arising from her adipose tissue stores, she will also be contributing newly ingested PCE associated with the dietary fats. This could be especially important in residential exposure to PCE where air contamination may be elevated due to proximity to a dry cleaning shop or other establishment where PCE is used in large quantities. In addition, the association of PCE with fats may greatly impact the workers in a dry cleaning shop where lunch is eaten on the premises, or left where the food may adsorb PCE from the air. There is very little information on the extent to which foods may be contaminated in residential dry cleaner settings, or in the dry cleaning facility itself.

It is interesting to recall the case reported by Bagnell and Ellenberger (1977) when an infant suffered liver damage, possibly as a result of being nursed by a mother whose milk was contaminated by PCE. The case report indicated that the mother often had lunch at her husband's place of work at a dry cleaning facility. Although the air levels of PCE were not available, the levels were sufficiently elevated to cause acute symptoms in the husband who was employed there and occasionally in his wife. It is possible that the mother's milk became contaminated not only via the inhalation of PCE from the air, but also from ingestion of PCE with the fats in her lunch, which subsequently transfer to breast milk.

Breast Milk

The lactating woman can be considered a potential reservoir for airborne PCE due to the inhalation of PCE and its eventual presence in the fat of her blood, breast milk, and adipose tissue. PCE contamination of breast milk is a secondary result of the mother's inhalation of contaminated air. If the mother/infant pair are both exposed to contaminated air in the residence or workplace, the air contamination exposure greatly exceeds the infant's exposure from ingestion of contaminated breast milk. Only limited sampling of breast milk for the presence of PCE and other VOCs has been conducted. Based on sampling of 17 nursing women (Sheldon et al., 1985), however, PCE appears to be a nearly ubiquitous contaminant in the breast milk of nonoccupationally exposed women. No sampling has been reported in breast milk samples of women who currently or previously were employed in the dry cleaning industry. The elevated airborne levels typically found in dry cleaning establishments coupled with the relatively long adipose tissue storage of unmetabolized PCE point to the great likelihood of significant concentrations of PCE that would be found in the milk of occupationally exposed women.

The nursing mothers study (Sheldon et al., 1985) was part of an ambitious series of studies conducted in the 1980s by the U.S. EPA called the Total Exposure Assessment Methodology (TEAM) Study. The TEAM Study was designed to develop and demonstrate methods of measuring human exposure to environmental pollutants through all important exposure routes (air, water, food) simultaneously and to measure the resulting body burden in breath, blood, and urine. The main TEAM study investigated the presence of selected volatile and semivolatile organic compounds in air, water, and exhaled breath. In addition, a special study was devoted to nursing mothers. The goal of the nursing mothers study was to determine the concentrations of selected toxic substances in the environmental samples, and in the blood, breath, urine, and mother's milk. The study participants were 17 lactating women identified from the larger group of about 300 randomly selected residences in the main TEAM study. The nursing mothers study provides the only human data that concurrently measure PCE in personal air, exhaled breath, blood, and milk of nonoccupationally exposed nursing mothers.

The Sheldon et al. (1985) findings are presented in Table 5-6. The results indicate that of the 17 nursing mothers studied, detectable levels of PCE were pervasive in the media sampled. All overnight and daytime personal air samples contained detectable levels of PCE, and 88% of exhaled breath, 22% of blood, and 63% of milk samples had detectable levels of PCE. The PCE concentrations in the milk samples ranged from nondetectable (less than 0.15 μg/L) to 43 μg/L. The limits of detection for air, blood, and milk samples were 1.8 $\mu g/m^3$, 0.4 μg/L, and 0.15 μg/L, respectively.

The nursing mothers in the study were compared to nonnursing females of reproductive age (16–40 years) measured for the same parameters, except milk, in the main TEAM study. For all prevalent volatile organic compounds, median personal air exposures were comparable or greater for the nursing mothers, but

Table 5-6   Tetrachloroethene Concentrations in Nursing Mothers Study

Medium	% Positive	Range[a]	Median	Arithmetic mean	75th Percentile
Overnight personal air, $\mu g/m^3$	100	1.1–210	11	28	22
Daytime personal air, $\mu g/m^3$	100	1.3–140	10	26	34
Exhaled breath, $\mu g/m^3$	88	1.0–85	4.6	13	5.7
Blood, μg/L	22	0.22–8.7	0.88	1.7	1.1
Milk, μg/L	63	0.15–43	1.4	6.2	3.8

Note: From Sheldon et al. (1985), n = 17 samples.
[a]Range includes not detected and trace values.

the median exhaled breath concentrations were lower usually by a factor of two. The authors suggest that the reduced concentrations of VOCs in exhaled breath of lactating women are due to greater retention of the lipophilic VOCs in milk fat. The greater reservoir of fats, including milk fat, in the nursing mother may accept a larger fraction of the inhaled VOCs delivered by the blood, thereby lowering the concentration in venous blood and exhaled breath (Wallace et al., 1989). Comparing the ratios of air exposure and breath levels for PCE, an estimated 50–80% of the PCE normally exhaled is retained by the nursing mother.

Spearman correlations between milk PCE and breath PCE concentrations were found to be significant ($R = .73$), as were milk PCE and daytime personal air PCE concentrations ($R = .75$). The relationship between milk PCE and blood PCE concentrations was not found to be significant ($R = .45$), probably due to the lack of sensitivity of blood PCE analysis at low concentrations. However, blood and breath PCE concentrations were highly correlated ($R = .81$). Unfortunately, individual values for PCE in overnight and daytime personal air samples were not provided in the report although they were measured. The results cannot be obtained at this time (L. Sheldon, personal communication, 1991).

PHARMACOKINETIC MODELING

A number of physiologically based pharmacokinetic (PBPK) models have been developed by different investigators to predict tissue concentrations based on exposure (Ward et al., 1988; Sielken, 1990; Hattis et al., 1990; Bogen & McKone, 1988; Travis et al., 1989). The PBPK models estimate the uptake, metabolism, distribution, storage, and elimination of chemicals based on physiological parameters and chemical-specific characteristics. Although PBPK models among investigators differ in parts of their mathematical treatments, their foundations are all based on a similar concept. That is, if one knows the appropriate metabolic and partitioning factors for a specific chemical, the expected levels of the chemical in body compartments can be predicted.

Figure 5-1 shows a diagram of a typical pharmacokinetic model used to simulate the apportioning of a chemical in the simplified human body. The body is seen as a series of compartments grouped on the basis of physiological parameters (blood flow rates, breathing rates, and tissue volumes). Chemical-specific parameters will determine how the chemical partitions between air and blood, and blood and tissues. Metabolism (which is assumed to occur only in the liver in most models) is described using both a linear metabolic component and a Michaelis–Menton component describing saturable metabolism.

PBPK models are useful to estimate concentrations, but are not meant to replace monitoring studies. Uncertainties and variation in the values assigned to various parameters (inhalation rate, blood and organ volumes, fat content, and body weight considerations) may lead to inaccurate or misleading assessments. However, in the absence of monitoring data, I used pharmacokinetic modeling

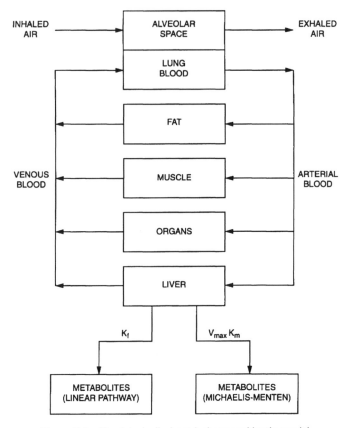

Figure 5-1 Physiologically based pharmacokinetic model.

and physicochemical relationships to predict the milk PCE concentrations for different exposure conditions. The assumptions made for these calculations are that the fat content of blood and milk is 0.65% (Task Group on Reference Man, 1974) and 4% (Lawrence, 1989), respectively, and that equilibrium has been reached between the air PCE concentration to which the woman has been exposed and the resultant tissue PCE concentrations.

Although none of the previously developed models is specifically designed to estimate breast milk concentrations, the predicted adipose tissue concentration can be used to estimate the breast milk concentration, assuming milk contains 4% fat. A limitation of this approach is the absence of a "milk compartment" in the PBPK model, resulting in predicted milk concentrations that do not include a mathematical treatment to account for the removal of milk from the mother's body. The significance of the lack of a "milk compartment" in the PBPK model is not known.

In this assessment, a commercially available PBPK model by Sielken (1990) was used to estimate breast milk PCE concentrations. The PBPK parameters used for this assessment are shown in Appendix 5-1. As shown in Table 5-7, the following maternal exposure scenarios were assessed:

A   Occupationally exposed mother inhaling air containing PCE at the American Conference of Governmental Industrial Hygienists (ACGIH) threshold limit value (TLV) of 340 mg/m³, 8 h/day, followed by exposure to an indoor residential background concentration of 27 µg/m³ (mean indoor air concentration at control residences identified in Schreiber et al., 1993).

B   Occupationally exposed mother inhaling air containing PCE at the OSHA permissible exposure limit (PEL) of 170 mg/m³, 8 h/day, followed by exposure to an indoor residential background concentration of 27 µg/m³.

C   Occupationally exposed mother inhaling air containing PCE of 40 mg/m³ (approximate arithmetic mean concentration for counter workers, pressers, and seamstresses) 8 h/day, followed by exposure to an indoor residential background concentration of 27 µg/m³.

D   Nonoccupationally exposed mother inhaling air containing PCE of 45.8 mg/m³ [24-h average concentration of PCE reported by Schreiber et al. (1993) in an apartment above a dry cleaner using an old dry-to-dry machine]

E   Nonoccupationally exposed mother inhaling air containing PCE of 7.7

Table 5-7   PBPK-Simulated Tetrachloroethene Concentrations in Biological Media

Mother's exposure scenario	Maximum simulated concentration (µg/L)			
	Blood	Fat	Milk[a]	Infant dose from milk[b] (mg/kg/day)
A. 8 h at 340 mg/m³, then 16 h at 27 µg/m³	1320	211,000	8440	0.82
B. 8 h at 170 mg/m³, then 16 h at 27 µg/m³	557	88,350	3530	0.34
C. 8 h at 40 mg/m³, then 16 h at 27 µg/m³	132	21,400	857	0.08
D. 24 h at 45.8 mg/m³	470	74,900	3000	0.3
E. 24 h at 7.7 mg/m³	79	12,600	500	0.05
F. 24 h at 250 µg/m³	2.6	400	16.2	0.0015
G. 24 h at 27 µg/m³	0.23	38	1.5	0.0001

Note: Sielken (1990) PBPK model results.
[a]4% of fat concentration.
[b]Assumes 7.2-kg infant ingests 700 ml breast milk/day.

mg/m³ [24-h average concentration reported by Schreiber et al. (1993) in apartments above dry cleaners using transfer machines]

F Nonoccupationally exposed mother inhaling air containing PCE of 250 µg/m³ [24-h average reported by Schreiber et al. (1993) in apartments above dry cleaners using dry-to-dry machines].

G Nonoccupationally exposed mother inhaling air containing PCE at an indoor residential background concentration of 27 µg/m³ (Sheldon et al., 1985).

Table 5-7 presents the results of the PBPK modeling of the seven exposure scenarios just given. The predicted breast milk PCE concentration for women exposed under occupational conditions ranged from 857 to 8440 µg/L. The predicted breast milk concentrations for women exposed to PCE in residences near dry cleaners was estimated to range from 16 to 3000 µg/L. Typical residential exposure resulted in a predicted breast milk PCE concentration of 1.5 µg/L. The estimated infant dose from breast milk was calculated assuming that a 7.2-kg infant ingests 700 ml breast milk/day.

One can also estimate milk PCE concentrations by use of the fat:blood partition coefficient, $\lambda(f/b)$, which relates the relative concentrations based on the solubility of PCE in fat and blood. A fat:blood partition coefficient of 159 for PCE was found experimentally (Bogen & McKone, 1988). The partition coefficient indicates that PCE will be present in fat at 159 times its presence in blood. Table 5-8 shows the predicted milk PCE concentrations based on the measured blood PCE concentrations using the fat:blood partition coefficient. The predicted milk PCE concentrations are quite similar to those derived using

Table 5-8 Tetrachloroethene Concentrations (µg/L) in Blood and Milk Based on Fat:Blood Partition Coefficient

Measured PCE in blood	PCE predicted in fat[a]	PCE in milk[b]		Reference
		Predicted	Measured	
8.7	1,383.3	55.3	43.0	Sheldon et al. (1985)
4.2	667.8	26.7	2.7	
2.9	461.1	18.4	1.4	
3.9	620.1	24.8	26	
3000	447,000	19,080	10,000	Bagnell and Ellenberger (1977)
1000[c]	159,000	6360	NA	ACGIH (1990)
2300[c]	365,700	14,628	NA	Koizumi (1989)
2500fl[d]	397,500	15,900	NA	Ohde and Bierod (1989)

Note: NA, not available.
[a]Blood concentration times $\lambda(f/b)$ = 159.
[b]Fat concentration times 4%.
[c]Blood concentration predicted after exposure to PCE at TLV of 50 ppm (344.5 mg/m³) for 8 h.
[d]fl, Blood concentration measured in neighbors of dry cleaning establishments.

the relative amounts of fat present in blood and milk. Both methods indicate the clearly elevated levels that are expected to result in both blood and milk of women exposed to PCE at the TLV. Also of note are the highly elevated blood PCE measurements made by Ohde and Bierod (1989) in samples obtained from nonoccupationally exposed neighbors of dry cleaning establishments. They measured blood PCE concentrations of 2500 µg/L, similar to the blood concentrations expected from occupationally exposed workers, and not much below the blood concentration of 3000 µg/L measured in the mother whose infant suffered liver damage after nursing exposure (Bagnell & Ellenberger, 1977).

Of the four nursing mothers with detectable blood PCE concentrations reported by Sheldon et al. (1985), the estimated milk PCE concentrations are very close to the measured values in two cases. In the two others, the predicted values are about an order of magnitude greater than the measured milk concentrations. The PBPK-estimated milk concentrations based on the blood PCE concentrations reported by Bagnell and Ellenberger (1977) are also within an order of magnitude of the measured concentrations. The agreement of the modeled milk PCE concentrations with the measured concentrations are an indication that the model appears to provide reasonable estimates of milk PCE levels at both high and low concentrations.

In summary, air and blood PCE concentrations as well as PBPK modeling can be used to predict PCE concentrations in breast milk. The predicted values are in generally good agreement with measured values, where available (see Tables 5-7 and 5-8). Exposure of the general public to PCE in air, using an arithmetic average air concentration of 27 µg/m³, results in a measured mean blood PCE concentration of 1.7 µg/L and milk PCE concentration of 6.2 µg/L (Sheldon et al., 1985). PBPK modeling using an air concentration of 27 µg/m³ predicts a blood concentration of 0.23 µg/L and a milk concentration of 1.5 µg/L. Monitoring studies should be conducted to validate the PBPK-predicted milk and blood concentrations, since no information is available for women exposed to PCE under these residential exposure conditions.

Comparisons to Standards and Guidelines

Standards and guidelines for PCE in drinking water have been established by the U.S. EPA. For noncarcinogenic effects, guideline concentrations were based on a no-observed-effect level of 20 mg/kg/day found by Buben and O'Flaherty (1985), who administered PCE by gavage to male Swiss-Cox mice 5 days/week for 6 weeks. Higher dose levels induced hepatic changes. The uncertainty factors used in the derivation of drinking water guidelines ranged from 100 to 1000 (IRIS, 1991).

The U.S. EPA (1991a) recently proposed maximum contaminant levels (MCLs) for drinking water. The MCL and MCLG (goal) are based on the assessment of PCE as a carcinogen. In its most recent reevaluation of the adverse effects of exposure to PCE, the U.S. EPA reviewed the evidence for the carcinogenicity of PCE for use in proposing maximum contaminant level goals (MCLGs) and national primary drinking water regulations (NPDWR) (U.S. EPA, 1991a). The EPA found strong evidence of carcinogenicity from ingestion of PCE based on consideration of the weight of evidence, pharmacokinetics, and exposure. The EPA assessment for carcinogenicity was based on the epidemiologic data and animal data as supplemented by metabolism information and results from short-term studies. Based on the EPA assessment, PCE was placed in Category I, for carcinogens classified as Group A, B1, or B2. As a matter of policy, the EPA sets MCLGs at zero for chemicals in Category I.

MCLGs are set at concentrations at which no known or anticipated health effects would occur, allowing for an adequate margin of safety. Establishment of a specific MCLG depends on the evidence of carcinogenicity from drinking water exposure or the EPA's reference dose (RfD). Since PCE is classified as Category I, carcinogen, the MCLG is based on its carcinogenic nature, and an RfD for drinking water is not calculated. An MCL of 0.005 mg/L (5 µg/L) was established based on best feasible control technology.

However, the U.S. EPA has established a reference dose (RfD) for PCE to evaluate chronic oral exposure, although it was not used to derive the drinking-water MCL or MCLG. The RfD (formerly known as an acceptable daily intake, ADI) is an estimate, with an uncertainty spanning perhaps an order of magnitude, of a daily exposure to the human population (including sensitive subgroups) that is likely to be without an appreciable risk of deleterious health effects (excluding carcinogenic effects) during a lifetime. The RfD is derived from a no or lowest observed adverse effect level (NOAEL or LOAEL) that has been identified from a subchronic or chronic scientific study of humans or animals. The NOAEL or LOAEL is then divided by the uncertainty factor to derive the RfD.

For PCE, the chronic oral RfD is 0.01 mg/kg/day (U.S. EPA, 1991b; IRIS, 1991). For carcinogens such as PCE, the oral Rfd is not used to calculate an MCLG. The oral RfD can be used, however, as a measure to assess risk of adverse health effects (other than cancer) due to ingestion of substances containing PCE (U.S. EPA, 1991b). The scenarios for occupational exposure all result in predicted breast milk concentrations that lead to infant exposure from 8 to 60 times the (adult) oral RfD. This indicates that the breast-fed infant is ingesting PCE (in mg/kg/day) at levels exceeding the recommended chronic ingestion limit for adults. The range of PCE intake via breast milk for the infant of the nonoccupationally exposed mother is well below the oral RfD, with the exception of a nursing mother living in a dwelling adjacent to a dry cleaning establishment.

There are no criteria available for PCE in food. Generally, nondetectable levels of PCE in foods are expected with the exceptions of fat-containing foods exposed to PCE via extraction processes or by contamination due to the vicinity of fugitive PCE emissions from dry cleaning shops, as previously discussed. Based on the scanty database available, breast milk will generally contain nondetectable to moderate levels in the milk of women in the general population. Highly elevated PCE levels in milk from women occupationally exposed are predicted based on exposure scenario and physiochemical parameters of its distribution in the mother.

RISKS AND BENEFITS OF BREAST MILK

It is difficult to weigh the potential adverse effects of exposure to contaminants in breast milk against the recognized benefits afforded by breast milk. On the one hand, the levels of PCE predicted in breast milk of heavily exposed women are predicted to result in breast milk PCE doses close to the levels at which adverse health effects have been demonstrated. These doses are also predicted to contribute significant excess cancer risks. These predicted risks are estimates that have a large degree of uncertainty associated with them. On the other hand, the benefits of breast milk are well known and have been measured by evaluating infant health. Furthermore, infants in residences in buildings with dry cleaners are exposed to PCE primarily from inhalation of contaminated air rather than from ingestion of contaminated breast milk.

Risk assessment is a useful tool to assess chemical risks, but has inherent uncertainties that may underestimate or overestimate risks. The uncertainties introduced in a risk assessment arise from the lack of appropriate human data for evaluation and the reliance on animal studies to predict effects of chemical exposure in humans. Throughout the risk assessment process, conservative assumptions may be made regarding a variety of factors that may result in overestimation of risk. Upper-bound assumptions may generate exaggerated estimates of overall risk, even if only moderate overestimates are made at each decision-making juncture. Assumptions that may lead to exaggerated risk include the amount of chemical absorption and retention, the use of data from the most sensitive animal species studied (which may overestimate human sensitivity), magnitude of safety factors used to derive noncarcinogenic reference dose (RfD) levels, and uncertainties in animal models used to predict excess cancer risk in humans based on animal data (Kimbrough, 1987). On the other hand, some assumptions may result in underestimation of risk for particular situations or populations, such as infant exposure under steady-state conditions for prolonged exposure periods. Infant dose, absorption, and metabolism may be significantly different than an adult's, and may result in underestimation of infant risks.

Especially in risk assessments that evaluate carcinogenicity, the animal models upon which the assessments are made have not been verified by data at the lower end of the dose-response curve. Criticism aimed at the risk assessment process usually centers on the lack of human studies or findings that are inconclusive or negative regarding the cancer incidence in human populations at risk despite positive findings in animal bioassays. Unfortunately, this dilemma is not easily rectified, as the human studies necessary are costly and time-consuming and have methodological drawbacks not easily overcome (Rogan et al., 1991).

Another source of uncertainty in risk assessment is the assumptions that are used to evaluate infant exposure. On average, a daily intake of 700 ml breast milk containing 4% lipid content is assumed to be ingested by a 7.2-kg infant for 1 year. While these standard assumptions are sound and based on citations from the scientific literature, individuals often vary from the quoted norm, since factors such as breast-feeding duration and infant growth rates are highly variable.

One of the most important uncertainties regarding infant exposure via breast milk is the lipid content of the milk. Since the contaminants in milk most likely to be present in large amounts are lipid soluble, the percent lipid is critical to determine the exposure of the infant. Monitoring data on breast milk contaminants usually present the contaminant information on a whole milk or milk fat basis. In either case, the milk fat (lipid) content allows an estimate of the contaminant ingestion by the infant. Many investigators do not present the fat content of the milk examined and therefore leave to others the assumptions of fat content. Percent fat in breast milk is cited by various investigators, for example, from a mean of 2.5% by Rogan and colleagues (1991), a mean of about 2.1 to 2.9% by Foman (1974), a mean of 4% in a review article in *Nutrition Reviews* (1989), and 4.2% by Lawrence (1989) in her book, *Breastfeeding: A Guide for the Medical Profession*. I have consistently used 4% milk fat in my calculations for fat content as this figure represents an upper but reasonable estimate that is consistent with the findings of others. Actual contaminant levels in milk and the attendant risks, therefore, may actually be lower.

On the other hand, the sensitivity of an infant to chemical insult may be greater than that of an adult due to immaturity of the infant's immune system, neurological development, and organ system maturation. Many of the animal studies that provide the basis for the reference dose and cancer potency factors are conducted on mature laboratory animals whose development may be different than a human newborn. Most "lifetime" animal experiments begin when the animal is 6 weeks old and has already been weaned by its mother (Rogan et al., 1991). It is generally believed that the young are more sensitive to the adverse effects of chemical exposure than the adults. Therefore, risk assessment may underestimate the actual impact of a chemical on the infant who is nursed with contaminated milk.

Breast milk provides a complete economical source of nourishment for infants through the age of 6 months (Grandjean et al., 1988; Lawrence, 1989; Cunningham et al., 1991). As the infant ages, additional foods are required to meet the infants' needs for growth and nutrition. The benefits of breast milk have been evaluated in terms of reduced rates of infant morbidity and mortality compared to formula-fed infants.

Quantitative assessments of the benefits of breast-feeding have estimated that breast-feeding provides greatly improved mortality and morbidity rates compared to those rates for formula-fed infants for all socioeconomic strata. In developed countries, reductions in mortality rates in breast-fed infants have been estimated to range from about 2560 to 5100 per million infants when compared with formula-fed infants (Cunningham et al., 1991; Rogan et al., 1991; Lucas & Cole, 1990; Madeley et al., 1986; Carpenter et al., 1983; Carpenter & Emery, 1974). The studies upon which these estimates are based do not consider the presence or absence of chemicals in breast milk or infant formula.

Rogan and colleagues (1991) are the only researchers who have published accounts of both quantitative benefit and risk estimates afforded by the infants' exposure to breast milk and typical levels of its contaminants. They conclude that only extreme levels of contaminants in breast milk represent more of a hazard than failure to breast-feed. Others, such as the World Health Organization (Grandjean et al., 1988) and La Leche League (Cahill, 1982), have made subjective qualitative assessments of the benefits of breast-feeding and have assumed that the risks associated with the milk contaminants are very small or negligible compared to the benefits. Other investigators have assessed the risks from the presence of selected chemicals on breast milk quality but have not assessed the benefits provided by breast milk (Poitrast et al., 1988; Travis et al., 1988; American Pharmacy, 1983; Lauwers & Woessner, 1990; Wolff, 1983).

Just as uncertainties exist in the evaluation of risks, there are uncertainties in the assessment of the benefits afforded by breast-feeding. Sources of uncertainty in the epidemiologic studies that evaluate breast-feeding versus infant formula include (1) the failure to control for confounding variables such as socioeconomic status and access to medical care, which may be related to both the choice of the method of feeding and the outcome measure, (2) nonblind assignment of feeding mode, (3) failure to determine whether an infant's health condition determined the feeding method or the feeding method preceded the health condition, (4) small samples with the potential for unspecified biases, and (5) the lack of precision in specifying the categories of feeding and health outcomes (Kovar et al., 1984).

The definition of breast-feeding varies among researchers. The definition may be restricted to babies who are breast-fed exclusively, or it may include babies who are formula-fed with only a single daily feeding of breast milk. In general, the greatest mortality and morbidity differences are found when exclusive breast-feeding is compared with exclusive bottle-feeding; moderate differences are found with supplemental bottle-feeding (Cunningham et al., 1991). The impact of the duration of breast-feeding (3 months, 6 months, 1 year) on infant mortality and morbidity has not been quantified, as data are not available to evaluate the effect (Rogan et al., 1991).

Other methodological problems involve associated variables such as socioeconomic and environmental factors. The impact of the greater motivation and care-giving indicated by the intention to breast-feed is difficult to evaluate (Rogan et al., 1991). In developed nations, breast-fed infants are likely to come from small, nonsmoking families with high educational levels, high socioeconomic status, and excellent sanitation and home care. Therefore, it is difficult to discern how much of the reduced mortality and morbidity are due to breast-feeding per se or the associated environmental factors. The problem has been overcome by the use of studies conducted in suburban middle and upper class groups and mixed urban populations. Cunningham et al. (1991) conclude that these studies show, unequivocally, that breast-feeding prevents gastrointestinal and lower respiratory tract disease, otitis media, bacteremia, and meningitis in these settings. Therefore, despite these methodological problems, the weight of the evidence is overwhelming: There is little doubt that breast-feeding promotes infant health.

Even though the assignment of benefits is somewhat uncertain, the lives saved are measurable, in that records of births and deaths provide the information needed to derive a numerical assessment. The assessment of risks, on the other hand, provides theoretical estimates of increased cancer incidence in a population as the infants grow to adulthood. During the expected 70 years of life after the exposure to breast-milk contaminants, people are exposed to a multitude of chemicals in food, in the home and workplace, and as a result of personal lifestyle choices (cigarettes, alcohol, and drugs). Therefore the impact of the breast-milk contaminants on the overall health of an individual becomes more difficult to evaluate later in life, when the impact of the contaminants would theoretically become evident.

What is an acceptable level of risk for contaminants in breast milk? Various federal agencies are responsible for promulgating regulations and standards to protect the public from exposure to environmental chemicals. One of the primary factors that is considered in the decision to regulate is a determination of the relative increases in cancer risk associated with exposure to a chemical. In a review of cancer risk management, Travis et al. (1987) found that although there is no universally accepted definition of acceptable risk, most agency decisions are based on a combination of assessing "de manifestis" risk (obvious risk), "de minimus" risk (trivial risk), and the cost-effectiveness of managing risks that fall between these two levels.

In assessing the risks associated with chemical contaminants such as tetrachloroethene in breast milk, the benefits of breast milk should be weighed in relation to the risks of chemical exposure via breast milk. In general, most instances of environmental contamination are without any direct benefits to the exposed population (although indirect benefits may include greater agricultural productivity, reduced numbers of pests, etc.) and thus the evaluation of the benefits is not usually part of the risk management process. Breast milk is clearly different, since the estimated risks of adverse or carcinogenic effects associated with infant exposure to environmental contaminants are balanced by the well-established benefits of reduced infant mortality and morbidity from exposure to breast milk.

Government regulators have developed some guidelines with regard to risk assessment and risk management. It has been suggested that risks can be assessed in the following manner (Comar, 1979):

1 Eliminate any risk that carries no benefit or is easily avoided.
2 Eliminate any large risk (about 1 in 10,000 per year or greater) that does not carry clearly overriding benefits.
3 Ignore for the time being any small risk unless it can be easily avoided.
4 Actively study risks falling between these two limits, with the view that the risk of taking any proposed action should be weighed against the risk of not taking action.

The management of risk, however, necessitates accounting for validity and uncertainty in risk estimates, levels of acceptable risk, voluntary versus involuntary risks, societal risk perception, and ethical questions concerning privacy and free choice.

Various governmental agencies and scientists have defined acceptable risk in different ways according to explicit or implicit assumptions. There is no universally accepted definition of acceptable risk. Mantel and Bryan (1961) defined "virtually safe" as a risk of 1 in 100 million for a lifetime. The U.S. FDA considers a risk of 1 in 1 million for a lifetime as an insignificant public health concern (U.S. FDA, 1977), although the Delaney Clause in the Food, Drug, and Cosmetic Act instructs the FDA to not allow any additive to food that is found to be carcinogenic in animals or humans (Lave, 1986). The statutes of the Federal Insecticide, Fungicide and Rodenticide Act (FIFRA) require that the risks be balanced with the benefits.

Women in the general population were reported by Sheldon et al. (1985) to have mean and maximum PCE breast milk concentrations of 6.2 and 43 µg/L. These concentrations are estimated to potentially contribute 0.4 and 2.9 additional cancer cases per million population, respectively. There is a high likelihood of detecting PCE in breast milk [60% of the samples reported by Sheldon et al. (1985) had detectable PCE concentrations], but the estimated cancer risks related to these exposures (ranging from 0.4 to 2.9 per million) pale compared to the benefits afforded the infant via breast milk.

The presence of PCE in the breast milk of women exposed to PCE under occupational exposure settings is estimated by PBPK modeling to result in breast milk concentrations of up to 8000 µg/L. The cancer risk associated with infant ingestion of milk containing this concentration of PCE is 600 per million. Although this level of risk is substantial, it is considerably lower than the estimated benefits of breast milk exposure (reduction in mortality rates in breast-fed infants are estimated to range from about 2560 to 5100 per million infants when compared with formula-fed infants).

Women exposed to PCE in selected residential environments near dry cleaners are estimated by PBPK modeling to have elevated levels of PCE in breast milk. The breast milk PCE concentrations as a result of maternal inhalation exposure are estimated to range from 16.2 to 3000 µg/L. The cancer risks associated with infant exposure to these contaminant concentrations in breast milk are estimated to range from 1.4 to 220 per million population, and, although substantial, are considerably less than the benefits associated with breast milk.

The noncarcinogenic risks related to these exposures, however, may be significant. For the occupational scenarios evaluated and for the residential exposures near dry cleaning establishments, an infant may be ingesting doses of PCE in breast milk that are within an order of magnitude of doses associated with noncarcinogenic adverse health effects. The case report of Bagnell and Ellenberger (1977) identified hepatomegaly and jaundice in an infant exposed to PCE solely via breast milk. A frank effect level (FEL) of 1.4 mg/kg/day based on this report can be derived.

In contrast to carcinogenic risks, these risks are less theoretical since they have been known to result in acute adverse effects in an infant and are consistent with effects observed in experimental animals. The FEL can be compared to the doses of 0.08–0.82 mg/kg/day associated with breast milk contamination as a result of maternal occupational exposure. The residentially exposed mother may have breast-milk PCE that provides an infant with a PCE dose ranging from 0.0015 to 0.3 mg/kg/day.

PCE is present in breast milk primarily as a result of the mother's inhalation of the chemical. Although small amounts may be ingested via the maternal diet, this route of exposure is small compared to the inhaled dose. The presence of volatile organic chemicals (VOCs) in breast milk results from contributions from adipose tissue stores (which are themselves a result of inhaled doses in equilibrium with blood and adipose tissue) and from inspired air. Whereas the concentrations of persistent chemicals such as PCBs and DDT cannot be controlled or reduced easily, the concentrations of PCE and other VOCs in milk can be modified by changes in the mother's work and/or residence.

The general population is exposed to low levels of PCE due to its widespread presence in indoor air. Unlike chemicals such as PCBs and DDT, maternal adipose tissue concentrations of PCE will be reduced via exhalation over a period of weeks once exposure has ceased. If the mother is exposed occupationally, the infant would be exposed to PCE via breast milk, but would be spared the inhalation exposure at the work site. If the mother can avoid or reduce her occupational exposure to PCE, the infant's exposure would be similarly reduced.

Exposure of the mother and infant in a residence contaminated with PCE due to its proximity to a dry cleaning establishment, however, presents a more serious exposure dilemma. New York State Department of Health data (Schreiber et al., 1993) indicate that indoor air in residences in the same building as a dry cleaner contain clearly elevated concentrations of PCE. Because the residents may spend up to 24 h/day in the residence, their total exposure to PCE may approach or exceed occupational exposure. The mother and infant are exposed primarily via the inhalation of contaminated air. In addition, the infant is exposed via ingestion of contaminated milk, which is estimated to provide about 4% of the child's total exposure (Schreiber, 1993).

Since the major route of exposure for the mother–infant pair is the inhalation of contaminated air, risks can be reduced by eliminating or reducing the levels of the contaminant in the indoor air environment. The contamination of milk is a secondary route of exposure and is essentially a result of the very large maternal exposure via the inhalation of contaminated air. Maternal and infant exposures can be reduced by removing the mother and infant from the contaminated environment, by removing the contaminant emission source from the residential environment, or by the implementation of engineering modifications on the emission source that result in reduction of airborne solvent concentrations.

For volatile solvents with characteristics similar to PCE, the primary route of infant exposure will be the inhalation of contaminated air rather than the ingestion of contaminated milk. The infant's greater magnitude of the exposure via air than via milk is due to the fact that an average infant inhales 1.2 m^3 (1200 L) of air per day while she or he ingests 700 ml milk. For example, in a residence where the airborne PCE concentration is 45,800 µg/m^3, a breast-milk contaminant concentration of 3000 µg/L is predicted by PBPK modeling of the mother's inhalation exposure. Therefore, the infant exposure to PCE is 55,000 µg/day via inhalation and 2100 µg/day via the ingestion of contaminated breast milk (see Table 5-9).

Women working in occupations where lipophilic volatile solvents are used are likely to have elevated adipose tissue and breast milk concentrations of the solvents predicated on the pharmacokinetic characteristics of the solvent and the degree of exposure of the mother. There are no studies available that examine the milk concentrations of chemicals in relation to the mother's occupational exposure. Residential exposure to volatile chemicals as a result of proximity of the residence to commercial or industrial sources of exposure has not been

Table 5-9 Estimates of Infant Tetrachloroethene Dose Via Inhalation of Residential Indoor Air and Ingestion of Breast Milk

Maternal exposure scenario[a]	Inhalation[b] dose, µg/day	Ingestion[c] dose, µg/day	Total[d] µg/kg/day
A. 340 mg/m^3	32	5900	820
B. 170 mg/m^3	32	2470	350
C. 40 mg/m^3	32	600	88
D. 45.8 mg/m^3	55,000	2100	7900
E. 7.7 mg/m^3	9200	350	1300
F. 0.25 mg/m^3	300	11	43
G. 0.028 mg/m^3	32	1	4.6

Note: PCE absorption and distribution via inhalation and ingestion may be different and may result in differences in toxic effects.
[a]See text for details.
[b]Infant inhales 1.2 m^3/day.
[c]Infant ingests 700 ml breast milk/day.
[d]Infant weight of 7.2 kg.

thoroughly evaluated. With the exception of the study conducted by Sheldon et al. (1985), there are no studies that examine the milk concentrations of chemicals in relation to the mother's residential exposure.

PCE has the potential for presence in breast milk in elevated concentrations due to the high concentrations encountered in the workplace and the lipophilic character of the chemical. In addition to PCE, there are other workplace chemicals that are likely to impact the quality of breast milk in exposed women. Women working in occupations that use large amounts of volatile solvents (such as painters, artists, machinists, chemists, and other trades and professions using solvents in poorly controlled work sites or in their own homes) are likely to have elevated tissue concentrations of the solvent to which they are exposed. In addition, residential indoor air exposures that result from proximity to industrial or commercial sources are likely to present similar exposure patterns to those found in residences located in the same building as a dry cleaner. Small businesses such as shoe repair shops, furniture refinishers, antique stores where furniture restoration is conducted, printing shops, and other solvent-utilizing industries located in residential areas may contribute to residential exposure, especially when located in the same building as residences.

For the residential exposure scenarios, the high breast milk levels predicted account for only a fraction of the infant's total exposure, and point out the serious nature of the inhalation exposure. Since the milk concentrations alone are estimated to represent increased cancer risk and provide little margin of exposure to doses associated with adverse noncarcinogenic health risks, the infant's total exposure (air and milk) is clearly of public health concern.

The sensitivity of an infant to chemical insult may be greater than that of an adult due to immaturity of the infant's immune system, neurological development, and organ system maturation. Many of the animal studies upon which the reference dose and cancer potency factors are based are conducted on mature laboratory animals whose development may be different than a human newborn. Most "lifetime" animal experiments begin when the animal is 6 weeks old and has already been weaned by its mother (Rogan et al., 1991). It is generally believed that the young are more sensitive to the adverse effects of chemical exposure than the adults. Therefore, the risk assessment may underestimate the actual impact of a chemical on the infant who is nursed with contaminated milk.

How does breast milk actually protect infants? Although many components of breast milk have been identified, the honest answer is that we do not fully know (Cunningham et al., 1991). Human milk contains immunologic components and other host defense factors that counteract the enteric and nonenteric pathogens that remain even in a hygienic environment. The activity of many such factors has been demonstrated in vitro, but the presence of such factors has seldom been correlated with prevention of illness in infants (Cunningham et al., 1991). The immunologically active protective components in human milk include the immunoglobulins IgA, IgG, and IgM; lactoferrin; lysozyme; interferon; monocytic phagocytic phagocytes and macrophages; neutrophils; B and

T lymphocytes; plasma cells; and vaccine viruses (Lawrence, 1989). The complexity of human milk, coupled with the extremely varied interactions among the mother, the infant, and the external environment, indicates that a simple answer to this question is not likely in the near future. Cunningham et al. (1991) conclude that at all levels—biochemical, immunologic, physiologic, behavioral, and epidemiologic—there is much to learn.

The benefits of breast-feeding are measurable, in that records of births and deaths provide the information needed to derive a numerical assessment of infant mortality and morbidity. The risks from breast milk contamination, on the other hand, are theoretical estimates of increased cancer incidence in a population as the infants grow to adulthood. During infancy they may be overshadowed by the risk of exposure to chemicals through other routes (i.e., inhalation of VOCs). Over the expected 70 years of life after the exposure to breast-milk contaminants, people are exposed to a multitude of chemicals in food, in the home and workplace, and as a result of personal lifestyle choices (cigarettes, alcohol, and drugs).

Ideally, providing uncontaminated air, water, and food including breast milk is the best choice to protect the public from unwanted chemical exposures and risks. Protection of the mother from hazardous exposures both in the workplace and at home is key to promoting maximum purity of breast milk.

REFERENCES

American Conference of Governmental Industrial Hygienists. 1990. 1990–1991 Threshold Limit Values for Chemical Substances and Physical Agents and Biological Exposure Indices. Cincinnati, OH: ACGIH.

American Pharmacy. 1983. The transfer of drugs and other chemicals into human breast milk. *Nutr. Suppl.* 23:29–36.

Atkinson, H. C., Begg, E. J., and Darlow, B. A. 1988. Drugs in human milk. Clinical pharmacokinetic considerations. *Clin. Pharmacokinet.* 14:217–240.

Bagnell, P. C., and Ellenberger, H. A. 1977. Obstructive jaundice due to a chlorinated hydrocarbon in breast milk. *Can. Med. J.* 117:1047–1048.

Bauman, D. E., and Currie, W. B. 1980. Partitioning of nutrients during pregnancy and lactation: A review of mechanisms involving homeostasis and homeorhesis. *J. Dairy Sci.* 63:1514–1529.

Berne, R., and Levy, M. 1981. *Cardiovascular physiology*, 4th ed. St. Louis, MO: C. V. Mosby.

Bogen, K., and McKone, T. 1988. Linking indoor air and pharmacokinetic models to assess tetrachloroethylene risk. *Risk Anal.* 8(4):509–520.

Borgstrom, B., and Patton, J. 1991. Luminal events in gastrointestinal lipid digestion. In *Handbook of physiology. Section 6: The gastrointestinal system*, vol. IV, *Intestinal absorption and secretions*, eds. S. Schultz, chap. 22. Bethesda, MD: American Physiological Society.

Buben, J., and O'Flaherty, E. 1985. Delineation of the role of metabolism in the hepatotoxicity of TCE and PCE: A dose-effect study. *Toxicol. Appl. Pharmacol.* 78:105–122.

Cahill, M. A. 1982. Environmental Contaminants in Breast Milk. Franklin Park, IL: La Leche League International. information sheet no. 30. March.

Carpenter, R., and Emery, J. 1974. Identification and follow-up of infants at risk of sudden death in infancy. *Nature* 250:729.

Carpenter, R., Gardner, A., Emery, J., et al. 1983. Prevention of unexpected infant death. *Lancet* 1:723–727.

Casey, C., and Hambidge, K. 1983. Nutritional aspects of human lactation. In *Lactation*, eds. M. C. Neville and M. R. Neifert, pp. 199–248. New York: Plenum Press.

Chappell, J. E., Clandinin, M. T., and Kearney-Volpe, C. 1985. Trans fatty acids in human milk lipids: Influence of maternal diet and weight loss. *Am. J. Clin. Nutr.* 42:49–56.

Comar, C. 1979. Risk: A pragmatic de minimus approach. *Science* 203:319.

Cunningham, A. S., Jelliffe, D., and Jelliffe, E. 1991. Breast-feeding and health in the 1980s: A global epidemiologic review. *J. Pediatr.* 118:659–666.

Droz, P., and Guillemin, M. 1986. Occupational exposure monitoring using breath analysis. *J. Occup. Med.* 28(8):593–602.

Entz, R., and Diachenko, G. 1988. Residues of volatile halocarbons in margarines. *Food Addit. Contam.* 5:267–276.

Fawell, J., and Hunt, S. 1988. Tetrachloroethylene. *Environmental toxicology: Organic pollutants*, chap. 4. Chichester: Ellis Horwood.

Field, C. J., Angel, A., and Clandinin, M. T. 1985. Relationship of diet and fatty acid composition of human adipose tissue structure and stored lipids. *Am. J. Clin. Nutr.* 42(6):1206–1220.

Fiserova-Bergerova, V. 1983. *Modeling of inhalation exposure to vapors: Uptake, distribution, and elimination*, vols. 1 and 2. Boca Raton, FL: CRC Press.

Foman, S. 1974. *Infant nutrition*. Philadelphia: W. B. Saunders.

Gallenberg, L., and Vodicnik, M. 1989. Transfer of persistent chemicals in milk. *Drug Metab. Rev.* 21(2):277–317.

Gardner, D. K. 1987. Drug passage into breast milk: Principles and concerns. *J. Pediatr. Perinatal Nutr.* 1:27–37.

Gargas, M., Burgess, R., Voisard, D., Cason, G., and Anderson, M. 1989. Partition coefficients of low molecular-weight volatile chemicals in various liquids and tissues. *Toxicol. Appl. Pharmacol.* 98:87–99.

Gartrell, M., Craun, J., Podrebarac, D., and Gunderson, E. 1985. Pesticides, selected elements, and other chemicals in infant and toddler total diet samples, October 1978–September 1979. *J. Assoc. Off. Anal. Chem.* 68:842–861.

Gartrell, M. J., Craun, J. C., Podrebarac, D. S., and Gunderson, E. L. 1986. Pesticides, selected elements, and other chemicals in infant and toodler total diet samples, October 1980–March 1982. *J. Assoc. Off. Anal. Chem.* 69:138–145.

Garza, C., Schanler, R. J., Butte, N. F., and Motil, K. J. 1987. Special properties of human milk. *Clin. Perinatol.* 14:11–32.

Grandjean, P., Kimbrough, R., Tarkowski, S., and Yrjanheikki, E. 1988. Assessment of Health Risks in Infants Associated with Exposure to PCBs, PCDDs and PCDFs in Breastmilk. Report of a WHO Working Group. Copenhagen: WHO Regional Office for Europe.

Grunder, F., and Moffitt, A., Jr. 1988. Routes of entry and excretion. In *Methods for biological monitoring*, eds. T. Kneip and J. Crable, pp. 7–14. Washington, DC: American Public Health Association.

Guthrie, H. A., and Bagby, R. S. 1989. *Introductory nutrition*, 7th ed. St. Louis, MO: Times Mirror/Mosby.

Hachey, D. L., Thomas, M. R., Emken, E. A., Garza, C., Brown-Booth, L., Adolof, R. O., and Klein, P. D. 1987. Human lactation: Maternal transfer of dietary triglycerides labeled with stable isotopes. *J. Lipid Res.* 28:1185–1192.

Hattis, D., White, P., Marmorstein, L., and Koch, P. 1990. Uncertainties in pharmacokinetic modeling for perchloroethylene. I. Comparison of model structure, parameters and predictions for low dose metabolism rates for models derived by different authors. *Risk Anal.* 10(3):449–458.

Hurley, L., and Lonnerdal, B. 1988. Trace elements in human milk. In *Biology of human milk*, ed. L. Hanson, pp. 75–94. Nestlé Nutrition Workshop Series, vol. 15. New York: Raven Press.

Illing, H., Mariscotti, S., and Smith, A. 1987. *Toxicity review: Tetrachloroethylene.* Health and Safety Executive. London: Her Majesty's Stationery Office.

Integrated Risk Information System. 1991. Tetrachloroethylene (CAS no. 127-18-4). Washington, DC: U.S. Environmental Protection Agency.

Jackson, D., Imong, S., Silprasert, A., et al. 1988. Circadian variation in fat concentration of breast milk in a rural northern Thai population. *Br. J. Nutr.* 59.

Jarabek, A., Menache, M., Overton, J., Dourson, M., and Miller, F. 1990. The U.S. Environmental Protection Agency's inhalation RfD methodology: Risk assessment for air toxics. *Toxicol. Ind. Health* 6(5):279–301.

Jensen, A. A. 1983. Chemical contaminants in human milk. *Residue Rev.* 89:1–128.

Jensen, A. A., and Slosach, S. A. 1991. *Chemical contaminants in human milk.* Boca Raton, FL: CRC Press.

Jensen, R. G. 1989. *The lipids of human milk.* Boca Raton, FL: CRC Press.

Kawauchi, T., and Nishiyama, K. 1989. Residual tetrachloroethylene in dry-cleaned clothes. *Environ. Res.* 48:296–301.

Kimbrough, R. D. 1987. Human health effects of polychlorinated biphenyls (PCBs) and polybrominated biphenyls (PBBs). *Annu. Rev. Pharmacol. Toxicol.* 27:87–111.

Knopp, R. H., et al. 1985. Effect of postpartum lactation on lipoprotein lipids and apoproteins. *J. Clin. Endocrinol. Metab.* 60:542–547.

Koizumi, A. 1989. Potential of physiologically based pharmacokinetics to amalgamate kinetic data of trichloroethylene and tetrachloroethylene obtained in rats and man. *Brit. J. Ind. Med.* 46:239–249.

Kovar, M. G., Serdula, M. K., Marks, J. S., and Fraser, D. W. 1984. Review of the epidemiologic evidence for an association between infant feeding and infant health. *Pediatrics* 74:615–638.

Lauwers, J., and Woessner, C. 1990. *Chemical agents and breast milk. A comprehensive list of drugs and other agents and their effects on the nursing infant.* Garden City Park. NY: Avery.

Lave, L. 1986. Approaches to risk management. A critique. In *Risk evaluation and management,* eds. V. Covello, J. Menkes, and J. Mumpower, pp. 461–489.

Lawrence, R. A. 1989. *Breastfeeding. A guide for the medical profession,* 3rd ed. St. Louis, MO: C. V. Mosby.

Lawrence, R. 1994. *Breastfeeding: A guide for the medical profession,* 4th ed. St. Louis, MO: C. V. Mosby.

Lucas, A., and Cole, T. 1990. Breastmilk and neonatal necrotizing enterocolitis. *Lancet* 336:1519–1523.

Lucas, R. M., Pierson, S. A., Myers, D. L., and Handy, R. W. 1981. RTI National Human Adipose Tissue Survey Quality Assurance Program Plan. Preliminary Draft. RTI/1864/21–11. Research Triangle Park, NC: U.S. Environmental Protection Agency.

Ludwig, H., Meister, M., Roberts, D., and Cox, C. 1983. Worker exposure to perchloroethylene in the commercial dry cleaning industry. *Am. Ind. Hyg. Assoc. J.* 44(8):600–605.

Macey, I., Nims, B., Brown, M., and Hunscher, H. 1931. Human milk studies. VII. Chemical analysis of milk representative of the entire first and last halves of the nursing period. *Am. J. Dis. Child.* 42:569.

Madeley, R., Hull, D., and Holland, T. 1986. Prevention of postneonatal mortality. *Arch. Dis. Child.* 61:459–463.

Mantel, N., and Bryan, W. 1961. Safety testing of carcinogenic agents. *J. Natl. Cancer Inst.* 27(2):455–470.

Materna, B. 1985. Occupational exposure to perchloroethylene in the dry cleaning industry. *Am Ind. Hyg. Assoc. J.* 46(5):268–273.

McConnell, G., Ferguson, D., and Pearmann, C. 1975. Chlorinated hydrocarbons and the environment. *Endeavour* 34(121):13–18.

Miller, L., and Uhler, A. 1988. Volatile halocarbons in butter: Elevated tetrachloroethylene levels in samples obtained in close proximity to dry cleaning establishments. *Bull. Environ. Contam. Toxicol.* 41:469–74.

Minson, D. J., Ludlow, M. M., and Troughton, J. H. 1975. Differences in natural carbon isotope ratios of milk and hair from cattle grazing tropical and temperate pastures. *Nature* 256:602.

Monster, A., Boersma, G., and Steenweg, H. 1979. Kinetics of tetrachloroethylene in volunteers. Influence of exposure concentration and work load. *Int. Arch. Occup. Environ. Health* 42:303–310.

Montes, A., Walden, C., Knopp, R., Cheung, M., Chapman, J., and Albers, J. 1984. Physiologic and Supraphysiologic Increases in lipoprotein lipids and apoproteins in late pregnancy and postpartum. Possible markers for the diagnosis of "prelipemia" arteriosclerosis. *Arteriosclerosis* 4:407–417.

Morgan, A., Black, A., and Belcher, D. 1970. The excretion in breath of some aliphatic halogenated hydrocarbons following administration by inhalation. *Ann. Occup. Hyg.* 13(4):219–233.

National Research Council. 1989. *Recommended dietary allowances*, 10th ed. Washington, DC: National Academy Press.

New York State Department of Health. 1991. Compilation of Octanol/Water Partition Coefficients from the Literature. Unpublished. Albany, NY: NYSDOH.

Nutrition Reviews. 1989. Transfer of isotopic dietary fatty acids into human milk. *Nutr. Rev.* 47:75–77.

Occupational Safety and Health Administration. 1989a. Industrial Exposure and Control Technologies for OSHA Regulated Hazardous Substances. March. Washington, DC: U.S. Department of Labor.

Ohde, G., and Bierod, K. 1989. Tetrachloroethylene exposure in the neighborhood of dry cleaners. *Off. Gesundh-Wes.* 51:626–628 (in German).

Poitrast, B. J., Keller, W. C., and Elves, R. G. 1988. Estimation of chemical hazards in breast milk. *Aviat. Space Environ. Med.* 59:A87–92.

Prentice, A., Prentice, A. M., and Whitehead, R. 1981. Breast milk fat concentrations of rural African women. I. Short-term variations within individuals. II. Long-term variations within a community. *Br. J. Nutr.* 45:483.

Radish, B., Luck, W., and Nau, H. 1987. Cadmium concentrations in milk and blood of smoking mothers. *Toxicol. Lett.* 36:147–152

Raiha, N. 1989. Milk protein quantity and quality and protein requirements during development. *Adv. Pediatr.* 36:347–368.

Reed, C. B. 1908. A study of the conditions that require the removal of the child from the breast. *Surg. Gynecol. Obstet.* May:514–527.

Rivera-Calimlim, L. 1987. The significance of drugs in breast milk. *Clin. Perinatol.* 14:51–70.

Rogan, W., and Gladen, B. 1991. PCBs, DDE and child development at 18 and 24 months. *Ann. Epidemiol.* 1(5):407–413.

Rogan, W. J., Gladen, B. C., McKinney, J. D., Carreras, N., Hardy, P., Thullen, J., Tingelstad, J., and Tully, M. 1986a. Polychlorinated biphenyls (PCBs) and dichlorodiphenyl dichloroethene (DDE) in human milk: Effects of maternal factors and previous lactation. *Am. J. Public Health* 76:172–177.

Rogan, W. J., Gladen, B. C., McKinney, J. D., Carreras, N., Hardy, P., Thullen, J., Tinglestad, J., and Tully, M. 1986b. Neonatal effects of transplacental exposure to PCBs and DDE. *J. Pediatr.* 109:335–341.

Rogan, W., Blanto, P., Portier, C., and Stallard, E. 1991. Should the presence of carcinogens in breast milk discourage breast feeding? *Regul. Toxicol. Pharmacol.*13:228–240.

Rosell, S., and Belfrage, E. 1979. Blood circulation in adipose tissue. *Physiol. Rev.* 59(4):1078–1104.

Sato, A., and Nakajima, T. 1979. A structure-activity relationship of some chlorinated hydrocarbons. *Arch. Environ. Health* 34:69–75.

Savage, E. P., Keefe, T. J., Tessari, J. D., Wheeler, H. W., Applehans, F. M., Goes, E. A., and Ford, S. A. 1981. National study of chlorinated hydrocarbon insecticide residues in human milk, USA. *Am. J. Epidemiol.* 113:413–422.

Schreiber, J. 1993. Predicted infant exposure to tetrachloroethene in human breast milk. *Risk Anal.* 13(5):515–524.

Schreiber, J., House, S., Prohonic, E., Smead, G., Hudson, C., Styk, M., and Lauber, J. 1993. An investigation of indoor air contamination in residences above dry cleaners. *Risk Anal.* 13:335–344.

Shah, J., and Singh, H. 1988. Distribution of volatile organic chemicals in outdoor and indoor air. A national VOC, data base. *Environ Sci. Technol.* 22(12):1381–1388.

Sheldon, L.S., Handy, R. W., Hartwell, T. D., Leininger, C., and Zelon, H. 1985. Human Exposure Assessment to Environmental Chemicals—Nursing Mothers Study. Research Triangle Park, NC: U.S. Environmental Protection Agency.

Sielken, R. 1990. PBPK Modeler Documentation. Bryan, TX: Sielken, Inc.

Steinberg, D. 1985. Regulation of lipid and lipoprotein metabolism. In *Best and Taylor's physiological basis of medical practice*, ed. J. B. West, 11th ed., pp. 805–817. Baltimore, MD: Williams & Wilkins.

Sun, J., Wolff, R., Kanapilly, G., and McClellan, R. 1984. Lung retention and metabolic fate of inhaled benzo(a)pyrene associated with diesel exhaust particles. *Toxicol. Appl. Pharmacol.* 73:48–59.

Tabacova, S. 1986. Maternal exposure to environmental chemicals. *NeuroToxicology* 7:421–440.

Task Group on Reference Man. 1974. Report of the Task Group on Reference Man. A Report Prepared by a Task Group of Committee 2 of the International Commission on Radiological Protection. Oxford: Pergamon Press.

Thomas, K., Pellizzari, E., Perritt, R., Nelson, W., and Wallace, L. 1989. New Jersey TEAM 1987 Study: Effect of the introduction of dry-cleaned clothes on tetrachloroethylene levels in indoor air, personal exposures and breath. *Proc. 1989 EPA/AEWMA Int. Symp. Measurement of Toxic and Related Air Pollutants*, Raleigh, NC, May.

Tichenor, B., Sparks, L., and Jackson, M. 1988. Evaluation of Perchloroethylene Emissions from Dry-Cleaned Fabrics. Final Report March-May. EPA/600/2-88/061. Research Triangle Park, NC: U.S. EPA.

Travis, C. C., and Hattemer-Frey, H. A. 1989. A perspective on dioxin emissions from municipal solid waste incinerators. *Risk Anal.* 9:91–97.

Travis, C. C., Richter, S. A., Crouch, E. A. C., Wilson, R., and Klema, E. D. 1987. Cancer risk management. A review of 132 federal regulatory decisions. *Environ. Sci. Technol.* 21:415–420.

Travis, C. C., Hattemer-Frey, H. A., and Arms, A. D. 1988. Relationship between dietary intake of organic chemicals and their concentrations in human adipose tissue and breast milk. *Arch. Environ. Contam. Toxicol.* 17:474–478.

Travis, C., White, R., and Arms, A. 1989. A physiologically based pharmacokinetic approach for assessing the cancer risk of tetrachloroethylene. In *The risk assessment of environmental and human health hazards: A textbook of case studies*, ed. D. J. Panstenbach, pp. 769–796. New York: John Wiley and Sons.

U.S. Environmental Protection Agency. 1986a. Broad Scan Analysis of the FY82 National Human Adipose Tissue Survey Specimens. Volume I. Executive Summary. Office of Toxic Substances, EPA-560/5-86-035. Washington, DC: U.S. EPA.

U.S. Environmental Protection Agency. 1986b. Broad Scan Analysis of the FY82 National Human Adipose Tissue Survey Specimens. Volume II. Volatile Organic Compounds. Office of Toxic Substances, EPA-560/5-86-036. Washington, DC: U.S. EPA.

U.S. Environmental Protection Agency. 1986c. Broad Scan Analysis of the FY82 National Human Adipose Tissue Survey Specimens. Semi-Volatile Organic Compounds. Office of Toxic Substances, EPA-560/5-86-037. Washington, DC: U.S. EPA.

U.S. Environmental Protection Agency. 1986d. Broad Scan Analysis of the FY82 National Human Adipose Tissue Survey Specimens. Volume IV. Polychlorinated Dibenzo-*p*-dioxins (PCDD) and Polychlorinated Dibenzofurans (PCDF). Office of Toxic Substances, EPA-560/5-86-038. Washington, DC: U.S. EPA.

U.S. Environmental Protection Agency. 1986e. Broad Scan Analysis of the FY82 National Human Adipose Tissue Survey Specimens. Volume V. Trace Elements. Office Substances, EPA-560/5-86-039. Washington, DC: U.S. EPA.

U.S. Environmental Protection Agency. 1987a. Characterization of HRGC/MS Unidentified Peaks from the Analysis of Human Adipose Tissue, Volume I. Technical Approach. Office of Toxic Substances, EPA-560/5-87-002A. Washington, DC: U.S. EPA.

U.S. Environmental Protection Agency. 1987b. Characterization of HRGC/MS Unidentified Peaks from the Analysis of Human Adipose Tissue. Volume II. Appendices. Office of Toxic Substances, EPA-560/5-87-002B. Washington, DC: U.S. EPA.

U.S. Environmental Protection Agency. 1989a. *Exposure factors handbook.* EPA/600/8-89/043. Final Report. Washington, DC: U.S. EPA.

U.S. Environmental Protection Agency. 1989b. Identification of SARA Compounds in Adipose Tissue. Office of Toxic Substances, EPA-560/5-89-003. Washington, DC: U.S. EPA.

U.S. Environmental Protection Agency. 1991a. 40 CFR Parts 141, 142, and 143. National primary drinking water regulations—Synthetic organic chemicals and inorganic chemicals; Monitoring for unregulated contaminants; National primary drinking water regulations implementation; National secondary drinking water regulations. *Fed. Reg.* 56(20):3526–3614.

U.S. Environmental Protection Agency. 1991b. Health Effects Assessment Summary Tables. Cancer Potency Factors. Washington, DC: U.S. EPA.

U.S. Food and Drug Administration. 1977. Food producing animals. *Fed. Reg.* 42(35):10412-10437.

Verberk, M., and Scheffers, T. 1980. Tetrachloroethylene in exhaled air of residents near dry-cleaning shops. *Environ. Res.* 21:432-437.

Wallace, L. A., Pellizzari, E. D., Sheldon, L., Hartwell, T. D., Zelon, H., Perritt, K., and Sparacino, C. 1989. Personal exposures and body burdens of 34 environmental pollutants measured for 17 New Jersey nursing mothers. Unpublished. Washington, DC: U.S. Environmental Protection Agency.

Wanner, M., Lehmann, E., Morel, J., and Christen, R. 1982. The transfer of perchloroethylene from animal feed to milk. *Mitt. Geb. Levenmittelunters. Hyg.* 73:82–87.

Ward, R., Travis, C., Hetrick, D., Anderson, M., and Gargas, M. 1988. Pharmacokinetics of tetrachloroethylene. *Toxicol. Appl. Pharmacol.* 93:108–117.

West, D., Prinz, W., and Greenwood, M. 1989. Regional changes in adipose tissue blood flow and metabolism in rats after a meal. *Am. J. Physiol.* 257(*Regul. Integr. Comp. Physiol.* 26)R711–R 716.

Wilson, G. F., MacKensie, D. D., and Brookes, I. M. 1988. Importance of body tissues as sources of nutrients for milk synthesis in the cow, using C^{13} as a marker. *Br. J. Nutr.* 60:605–617.

Witschi, H., and Brain, J. 1985. *Toxicology of inhaled materials. General principles of inhalation toxicology.* Berlin: Springer-Verlag.

Wolff, M. S. 1983. Occupationally derived chemicals in breast milk. *Am. J. Ind. Med.* 4:259–281.

Wong, W., Butte, N., O'Brian Smith, E., Garza, C., and Klein, P. 1989. Body composition of lactating women determined by anthropometry and deuterium dilution. *Br. J. Nutr.* 61:25–33.

Woodward, D. R., Rees, B., and Boon, J. A. 1989. Human milk fat content: Within-feed variation. *Early Hum. Dev.* 19:39–46.

World Health Organization. 1984. Environmental Health Criteria 31. Tetrachloroethylene. Geneva: WHO.

World Health Organization. 1986. Environmental Health Criteria 59. Principles for Evaluating Health Risks from Chemicals During Infancy and Early Childhood: The Need for a Special Approach. Geneva: WHO.

Zabik, M., Hoojjat, P., and Weaver, C. 1979. Polychlorinated biphenyls, dieldrin and DDT in lake trout cooked by broiling, roasting or microwave. *Bull. Environ. Contam. Toxicol.* 21:136–143.

Appendix 5-1 Pharmacokinetic Parameters for PCE

	Parameter	Woman
Body weight (kg)	W	60
Alveolar ventilation (ml/min)	V_A	5.25×10^3
Total blood flow rate (ml/min)	Q_T	5.52×10^3
Blood flow to tissue (%Q_T)		
Muscle	Q_m	19
Liver	Q_l	25
Fat	Q_f	5
Kidney (vessel rich)	Q_k	51
Tissue volume (%W)		
Muscle	V_m	62
Liver	V_l	4
Fat	V_f	20
Kidney (vessel rich)	V_k	5
Lung	V_a	1.1
Partitioning coefficients		
Blood:air	λ_b	10.3
Muscle:blood	λ_m	7.77
Liver:blood	λ_l	6.82
Fat:blood	λ_f	159
Kidney:blood	λ_k	6.82
V_{max} (ng/ml/min)	V_{max}/V_l	24.3
K_m (ng/ml)	$K_m \lambda_l$	2046

Note: Adapted from Ward et al. (1988), except V_A and Q_T scaled by $(W)^{0.75}$.

Chapter 6

Solvent Abuse
and Developmental Toxicity

Georgianne Arnold

Deliberate inhalation of organic solvents is a popular form of drug abuse that may have serious consequences for both abusing pregnant women and their unborn children. Organic solvents are aromatic compounds present in paint, glue, paint thinner, gasoline, and aerosol propellants and are utilized extensively in industry. They are properly classified as anesthetics and typically produce a temporary period of stimulation before central nervous system depression. When deliberately concentrated and inhaled these highly volatile compounds are rapidly absorbed in the lungs and lead to a rapid, brief "high." Their widespread availability and relatively low cost make them popular drugs of choice in certain populations. Although many solvents such as benzene, butane, and others are widespread, up to 99% of the solvent present in the most frequently abused substances (paint and paint thinners) is toluene (Donald et al., 1991). In addition, toluene is found in gasoline, dyes, stain removers, polishes, and other common occupational and consumer substances.

Organic solvents have a high affinity for lipid-rich tissues including the brain and central nervous system. The physiologic effects of solvent intoxication are well described and have been reported at concentrations of 500 ppm (Low et al., 1988). Acute effects include dizziness, slurred speech, euphoria, depressed respirations, tachycardia, and ultimately sleep. Chronic effects are serious and may be life-threatening. These include renal tubular acidosis, hypokalemia, pulmonary hypertension, restrictive lung disease, encephalopathy, peripheral neuropathy, and physical and mental dependence. The current permissible exposure for toluene in air according to the U.S. Occupational Safety and Health Administration is 100 ppm, well below the level of intoxication.

Solvent abuse was once thought to be the domain of adolescent boys. However, recent evidence points to a shift toward increasing adult use (Hershey & Miller, 1982). At Denver General Hospital, 7.5% of adult admissions for drug abuse were for solvent abuse (Streicher et al., 1981). Solvent abuse occurs in all ethnic and socioeconomic groups. However, there is some evidence that this practice is more prevalent in Mexican-American and Native American populations and among ethanol abusers (Arnold et al., 1994; Padilla et al., 1979; Pearson et al., 1994). Prevalence of abuse among women is also rising; some studies estimate up to half of all solvent abusers are women (Padilla et al., 1979; Davies et al., 1985).

FETAL EFFECTS

The potential teratogenicity of this rising drug of abuse is a cause for concern. In humans the half-life of toluene in blood is 3.4 h (Low et al., 1988). Therefore concentrated doses may remain in circulation for many hours. In addition, the main metabolic pathway of toluene (conversion to hippuric acid) does not appear to be functional in the fetus and newborn (Goodwin, 1988). Toluene crosses the placenta and has been detected up to 24 h after exposure in mouse fetal–placental compartments (Ghantous & Danielsson, 1986). Animal studies have shown decreased fetal weight, embryolethality, and cleft palate in toluene-exposed laboratory animals (Barlow & Sullivan, 1982; Nawrot & Staples, 1979; Hudak & Ungvary, 1978). The fetotoxic effect may be time and dose dependent; measurable reduction in fetal weight was seen after prolonged exposure to toluene at 133 ppm on days 6–13 of gestation in mice (Hudak & Ungvary, 1978), while doses of greater than 500 ppm were required in other studies (International Research and Development Corporation, 1985). In some cases fetal effects were seen without maternal toxicity. Animal studies have been reviewed in detail by Donald et al. (1991).

MATERNAL OCCUPATIONAL EXPOSURE

A number of studies have attempted to determine if women with occupational exposure to organic solvents have an increased risk of fetal anomalies. Case-control studies in Finland identified an increased risk for central nervous system malformations and oral clefts in infants of mothers with occupational solvent exposures (Holmberg, 1979; Holmberg & Nurminen, 1980; Holmberg et al., 1982). McDonald et al. (1987) identified an excess of solvent exposure in cases with renal-urinary or gastrointestinal defects compared to controls. He found most of the excess cases were associated with toluene exposure. Syrovadko (1977) found infants born to mothers with occupational toluene exposure from varnishes had decreased birth weight and fetal asphyxia. Two studies have noted a significantly increased risk of miscarriage in these women, one in which the average exposure was quantitated at 88 ppm (Ng et al., 1992; Huang, 1991). In the occupational exposure studies the solvent most consistently associated with teratogenicity was toluene. Other studies have not identified a relationship between occupational maternal solvent exposure and birth defects (Axellson et al., 1983; Olsen, 1983). Interpretation of occupational studies is complicated by the subjects' exposure to a variety of solvents, as well as variability in timing and poor documentation of extent of exposure. The effects of deliberate concentration and inhalation were not addressed in these studies.

MATERNAL RECREATIONAL ABUSE

Recreational abuse of toluene has been clearly established to cause maternal complications of pregnancy. Life-threatening renal tubular acidosis is a well-known complication associated with toluene sniffing (Taher et al., 1974), and dangerous maternal as well as fetal acidosis has been frequently identified among toluene-exposed pregnancies (Goodwin, 1988; Wilkins-Haug & Gabow, 1991; Lindemann, 1991). Acute fatty liver of pregnancy has also been reported (Paraf et al., 1993). The courses of 26 toluene-exposed pregnancies have been reported in detail (Table 6-1) (Wilkins-Haug & Gabow, 1991; Goodwin, 1991). In addition to renal tubular acidosis, other serious maternal complications have included premature labor, placental abruption, premature rupture of membranes, hypokalemia (including electrocardiographic changes), rhabdomyolysis, and occasional withdrawal.

A characteristic pattern of fetal anomalies is now emerging as a consequence of prenatal toluene exposure. The first report of the effect of solvents on the developing fetus appeared in 1979 (Toutant & Lippman, 1979) and described an infant with features suggestive of fetal alcohol syndrome born to a woman with chronic solvent abuse. To date, detailed outcomes of 63 births with prenatal exposure to toluene abuse have been reported (Hersh et al., 1985; Hersh,

Table 6-1   Pregnancy Complications in 26 Toluene-Exposed Pregnancies

Finding	Wilkins-Haug and Gabow (1991)	Goodwin (1988)
Renal tubular acidosis	10/21	5/5
Premature labor	18/21	3/5
Premature delivery	9/21	3/5
Premature rupture of membranes	2/21	—
Hypokalemia	10/21	3/5
Hypokalemia with EKG change	2/21	—
Placental abruption	2/21	—
Rhabdomyolysis	2/21	—
Withdrawal	4/21	—
History of alcohol abuse	3/21	—
Positive alcohol screen	1/21	—

1989; Goodwin, 1988; Arnold et al., 1994; Pearson et al., 1994). These cases
have overwhelmingly been associated with concentration and inhalation of spray
paint. They are associated with a high incidence of fetal death, prematurity,
neonatal and childhood growth retardation, microcephaly, and developmental
and speech delay (Table 6-2). This pattern of findings is reminiscent of early
reports of fetal effects of alcohol exposure in which increased incidence of poor
growth, malformations, and developmental delay in children of alcoholic mothers was noted (Jones et al., 1974).

In addition, 29 of these children have undergone a detailed dysmorphology
examination, leading to preliminary identification of a characteristic pattern of
craniofacial and limb defects known as the toluene embryopathy (Table 6-3)
(Hersh et al., 1985; Hersh, 1989; Arnold et al., 1994; Pearson et al., 1994).
Specifically, microcephaly, short palpebral fissures, abnormal auricles, small
jaw, thin upper lip, flat philtrum, narrow bifrontal diameter, abnormal scalp hair
patterning, hypoplastic fingernails, aberrant palmar creases, abnormal tone, developmental delay (including speech), and renal anomalies are common findings. Other frequently occurring findings include strabismus, downturned cor-
ners of the mouth, café au lait spots, and hemangiomas.

The similarities between these facial features and those of fetal alcohol
syndrome are striking. The roles of chronic acidosis, poor nutrition, and other
socioeconomic factors remain to be detailed in this population of children and
their mothers. However, chronic maternal acidosis of other causes has not resulted in this specific pattern of findings. The role of other contaminants introduced with the solvent must also be addressed. In addition, the specific duration
and level of exposure in these cases have not been carefully quantitated; however, most mothers had a one to two cans per day habit of spray paint and were
reportedly chronically toluene intoxicated. In these reported cases, a limited
number of mothers admitted to concurrent alcohol use. However, evidence of
alcohol use was rare although evidence of toluene abuse was striking. There-

Table 6-2 Outcome of 63 Births with Prenatal Toluene Exposure

Finding	Arnold et al. (1994)	Goodwin (1988)	Hersh et al. (1985), Hersh (1989)	Pearson et al. (1994)	All studies
Fetal death	3/35	0/5	0/5	1/18	6%
Prematurity	13/31	3/5	1/5	7/18	41%
Low birth weight	16/31	3/5	3/5	6/18	47%
Low birth length	6/31	—	1/4	—	20%
Birth microcephaly	7/31	—	2/3	5/18	27%
Weight <5%	11/24	—	3/5	3/9	45%
Length <5%	9/34	—	1/5	—	26%
Microcephaly	11/24	—	4/5	8/9	61%
Developmental delay	9/24	—	5/5	5/6	54%
Speech delay	9/24	—	5/5	—	48%

Table 6.3 Features of Toluene Embryopathy

Finding	Hersh et al. (1985), Hersh (1989)	Arnold et al. (1994)	Pearson et al. (1994)	All studies
Microcephaly	3/5	4/6	8/9	75%
Abnormal scalp hair pattern	2/5	0/6	8/18	34%
Narrow bifrontal diameter	4/5	3/6	6/18	45%
Strabismus	4/5	2/6	—	55%
Short palpebral fissures	5/5	5/6	10/18	69%
Deep-set eyes	5/5	0/6	—	45%
Epicanthic folds	1/5	0/6	—	9%
Small/hypoplastic midface	5/5	5/6	—	91%
Anomalous ears	5/5	4/6	8/18	59%
Flat/wide nasal bridge	4/5	5/6	—	82%
Small nose	4/5	3/6	4/18	38%
Flat philtrum	2/4	1/6	4/18	25%
Thin upper lip	1/5	2/6	9/18	41%
Small jaw	5/5	3/6	10/18	62%
Blunt fingertips	5/5	3/6	—	73%
Small nails	4/5	2/6	5/18	38%
Abnormal palmar crease	3/5	4/6	5/18	41%
Cryptorchidism	1/2	0/1	—	33%
Abnormal tone	5/5	0/6	3/18	28%
Hyperreflexia	3/5	1/6	—	36%
Attention deficit	4/5	1/6	—	45%
Renal anomaly (by ultrasound)	4/5	—	4/13	44%
Developmental delay	—	2/6	5/6	58%
Speech delay	—	1/6	—	17%

fore, alcohol abuse cannot account for the congenital anomalies found in these cases. Rather, the sharing of a common mechanism of teratogenicity between toluene and ethanol has been proposed (Pearson et al., 1994). In particular, the midfacial structures are derived from migrating mesodermal cells (Moore, 1982). Administration of ethanol to mouse embryo cultures has resulted in increased cell death of migrating mesoderm and neuroepithelial structures, giving rise to the classic features of fetal alcohol syndrome (Sulik & Johnston, 1983). In addition, ethanol induces cellular death in the region of the developing mouse mesonephric duct, consistent with the renal anomalies seen in fetal alcohol syndrome (Gage & Sulik, 1991). Animal studies will be required to determine if the facial and renal defects in toluene embryopathy share a similar pathophysiology. The pathophysiology of the limb defects remains to be determined.

Although chronic abuse by concentration and inhalation of toluene is likely to produce the most profound fetal effects, fetal toxicity from nondeliberate inhalation of toluene remains a possibility. The effect of short-term exposure from household activities such as painting may be significant if exposure is prolonged, occurs at a critical gestational timing, or is sufficient to produce maternal toxicity. It is likely that maternal and fetal factors mitigate susceptibility to this teratogen. Additional studies will be required to determine if a critical threshold of exposure or timing exists for toluene. This author is aware of one case of apparent toluene embryopathy in the child of a mother with extensive occupational toluene exposure during pregnancy that resulted in chronic mild maternal intoxication. Worker protection was provided by a paper mask.

CONCLUSIONS

Clearly, the identification of inhalant abuse is as critical as that of alcohol abuse with respect to the developing fetus. The clinician needs to be alert for telltale evidence of abuse including the appearance of intoxication, the odor of paint or glue, or paint staining around the mouth or cuticles. Toluene may not be detected on routine drug screens; however, it may be quantitated in urine and amniotic fluid when specifically requested. Addicted users may "huff" one or more cans per day, preferably of metallic spray paint. Factors such as (1) low cost, (2) easy availability, (3) increasing frequency of use in adults and women, (4) lack of popular knowledge of the teratogenicity of this drug habit, and (5) a dearth of treatment facilities for drug-abusing pregnant women may result in a new problem of social significance. We anticipate that continued recognition of toluene-exposed pregnancies will lead to more definitive description of the toluene embryopathy.

REFERENCES

Arnold, G. L., Kirby, R. S., Langendoerfer, S., and Wilkins-Haug, L. 1994. Toluene embryopathy: Clinical delineation and developmental follow-up. *Pediatrics* 93:216–220.
Axellson, O., Edling, C., and Andersson, L. 1983. Pregnancy outcome among women in a Swedish rubber plant. *Scand. J. Work Environ Health* 9:79–83.
Barlow, S. M., and Sullivan, F. M. 1982. *Reproductive hazards of industrial chemicals: An evaluation of animal and human data.* London: Academic Press.
Davies, B., Thorley, A., and O'Connor, D. 1985. Progression of addiction careers in young adult solvent misusers. *Br. Med. J.* 290:109–110.
Donald, J. M., Hooper, K., and Hopenhayn-Rich, C. 1991. Reproductive and developmental toxicity of toluene: A review. *Environ. Health Perspect.* 94:237–244.
Gage, J. C., and Sulik, K. K. 1991. Pathogenesis of ethanol induced hydronephrosis and hydroureter as demonstrated following in vivo exposure of mouse embryos. *Teratology* 44:299–312.
Ghantous, H., and Danielsson, B. R. G. 1986. Placental transfer and distribution of toluene, xylene, and benzene and their metabolites during gestation in mice. *Biol. Res. Pregnancy* 7:98–105.
Goodwin, T. 1988. Toluene abuse and renal tubular acidosis in pregnancy. *Obstet. Gynecol.* 71: 715–718.
Hersh, J. H., Podruch, P. E., Rogers, G., and Weisskopf, B. 1985. Toluene embryopathy. *J. Pediatr* 106:922–927.
Hersh, J. H. 1989. Toluene Embryopathy: Two new cases. *J. Med. Genet.* 26: 333–337.
Hershey, C. O., and Miller, S. 1982. Solvent abuse: A shift to adults. *Int. J. Addict.* 17: 1085–1089.

Holmberg, P. C. 1979. CNS defects in children born to mothers exposed to organic solvents during pregnancy. *Lancet* 2 :177–179.

Holmberg, P. C., and Nurminen, M. 1979. Congenital defects of CNS and occupational factors during pregnancy. A case referant-study. *Am. J. Ind. Med.* 1:167–176.

Holmberg, P. C., Hernberg, S., Kurppa, K., Riala, R., and Rantala, K. 1982. Oral clefts and organic solvent exposure during pregnancy. *Int. Arch. Occup. Environ. Health* 50: 371–376.

Huang, X. Y. 1991. Influence on benzene and toluene to reproductive function of female workers in leathershoe-making industry. *Chung-Hua Yu Fang I Hsueh Tsa Chih* 25:89–91.

Hudak, A., and Ungvary, G. 1978. Embryotoxic effects of benzene and its methyl derivatives: Toluene, xylene. *Toxicology* 11:55–63.

International Research and Development Corporation. 1985. Two-Generation Inhalation Reproduction/Fertility Study on a Petroleum Derived Hydrocarbon with Toluene. API Medical Research Publication no. 32-32854. Washington, DC: American Petroleum Institute.

Jones, K. L., Smith, D. W., Streissguth, A. P., and Myrianthopoulos, N. C. 1974. Outcome in offspring of chronic alcoholic women. *Lancet* 1:1076–1078.

Lindemann, R. 1991. Congenital renal tubular dysfunction associated with maternal sniffing of organic solvents. *Acta Paediatr. Scand.* 80:882–884.

Low, L. K., Meeks, J. R., and Mackerer, C. R. 1988. Health effects of the alkylbenzenes. I. Toluene. *Toxicol. Ind. Health* 4:49–75.

McDonald, J. C., Lavoie, J., Cote, J., and McDonald, A. D. 1987. Chemical exposures at work in early pregnancy and congenital defect: A case-referent study. *Br. J. Ind. Med.* 44:527–533.

Moore, K. L. 1982. *The developing human*, 3rd ed., pp. 197–200. Philadelphia: W. B. Saunders.

Nawrot, P. S., and Staples, R. E. 1979. Embryofetal toxicity and teratogenicity of benzene and toluene in the mouse. *Teratology* 19:41A.

Ng, T. P., Foo, S. C., and Yoong, T. 1992. Risk of spontaneous abortion in workers exposed to toluene. *Br. J. Ind. Med.* 49:804–808.

Olsen, J. 1983. Risk of exposure to teratogens amongst laboratory staff and painters. *Dan. Med. Bull.* 30:24-28.

Padilla, E. R., Padilla, A. M., Morales, A., Olmedo, E. L., and Ramirez, R. 1979. Inhalant, marijuana, and alcohol abuse among barrio children and adolescents. *Int. J. Addict.* 14:945–964.

Paraf, F., Lewis, J., and Jothy, S. 1993. Acute fatty liver of pregnancy after exposure to toluene. A case report. *J. Clin. Gastroenterol.* 17:163–165.

Pearson, M. A., Hoyme, H. E., Seaver, L. H., and Rimsza, M. E. 1994. Toluene embryopathy: Delineation of the phenotype and comparison with fetal alcohol syndrome. *Pediatrics* 93:211–215.

Streicher, H. Z., Gabow, P. A., Moss, A. H., Kono, D., and Kaehny, W. D. 1981. Syndromes of toluene sniffing in adults. *Ann. Int. Med.* 94:758–761.

Sulik, K. K., and Johnston, M. C. 1983. Sequence of developmental alterations following acute ethanol exposure in mice; Craniofacial features of the fetal alcohol syndrome. *Am. J. Anat.* 166:257–269.

Syrovadko, O. 1977. Working conditions and healthy status of women handling organsiliceous varnishes containing toluene. *Gig. Tr. Prof. Zabol.* 4:15.

Taher, S. M., Anderson, R. J., McCartney, R., Popovtzer, M. M., and Schrier, R. W. 1974. Renal tubular acidosis associated with toluene "sniffing." *N. Engl. J. Med.* 290:765–768.

Toutant, C., and Lippman, S. 1979. Fetal solvents syndrome. *Lancet* 1:1356.

Wilkins-Haug, L., and Gabow, P. A. 1991. Toluene abuse during pregnancy: Obstetric complications and perinatal outcomes. *Obstet. Gynecol.* 77:505–509.

Chapter 7

Breast Silicone Implants and Pediatric Considerations

Jeremiah J. Levine

Silicone breast implants have been in use for more than 20 years but the association between implants and diseases in women remains a source of great controversy. The problems attributed to breast implants include:

1 Local complications such as gel bleed or rupture, capsule formation and contracture, and skin breakdown
2 A possible association between implants and the subsequent development of breast cancer
3 The development of systemic illnesses including classic and atypical autoimmune diseases

Although many of these problems are not likely to affect children born to mothers with silicone implants, concern has been raised about possible second-generation effects, especially regarding autoimmune diseases. Little is known about the effects of silicone and silicone implants on children; however, several recent and slightly older studies suggest that there is an urgent need to investigate the issue of second generation effects.

SILICONE CHEMISTRY

Silicon is the second most abundant element found in our environment (Carlisle, 1984). Silicone is a synthetic compound containing a silicon–oxygen backbone (Le Vier et al., 1993). When organic (carbon-based) groups are attached to silicon, organosilicates are produced (Le Vier et al., 1993; Dunn et al., 1992). Generally, the carbon-to-silicon linkage in most commercial medical silicone products consists of methyl groups. The most commonly used compound for implants is polydimethylsiloxane (Berlin, 1994). Organic groups other than methyl may be substituted for methyl groups to alter the physical characteristics of the polymer. Depending on the degree of chain length and type of cross-linkage, an organosiloxane may be fluid, gel, or rubber. Silicone elastomer is produced by even greater cross-linkage (by substituting vinyl or hexenyl for methyl groups) together with the addition of amorphous silica (Yoshida et al., 1994). The most widely used breast implants, until their use was restricted by the U.S. Food and Drug Administration, have been silicone gel implants surrounded by the silicone elastomer (Kessler, 1992; Angell, 1992).

It is widely accepted that silicone gel can "leak" through the silicone envelope (Robinson et al., 1995; Winding et al., 1988; Pfleiderer et al., 1993). The formation of a fibrous capsule around the implant is thought to be a reaction to this material (Yoshida et al., 1994). Histologic evaluation of resected capsules revealed fibrous connective tissue together with foci of lipid-laden macrophages (Silver et al., 1993; Domanskis & Owsley, 1976; Gayou, 1979). In addition, retractile, nonpolarizable material is often found within the macrophages (Silver et al., 1993). Occasionally, multinucleated giant cells are also seen (Silver et al., 1993; Bridges & Vasey, 1993). Wells et al. (1994) demonstrated increased amounts of interleukin-6 in the capsule of patients with implants compared to control breast tissue and showed that the interleukin-6 was associated with macrophages. Together, these data suggest that "leaking" silicone can induce a local inflammatory reaction and may activate immune mediators.

In addition to local leakage, several studies have suggested that silicone may migrate to areas far removed from the breast tissue. Silicone-like material, often within macrophages, has been demonstrated in lymph nodes (Varga et al., 1989), skin (Silver et al., 1993; Teuber et al., 1995), and peripheral nerves (Sanger et al., 1992). Distal migration of silicone to the chest and abdomen has been documented (Silver et al., 1993). Using nuclear magnetic resonance, Pfleiderer et al. (1993) were able to demonstrate migration of silicone to the liver and spleen in rats following implantation.

POTENTIAL FOR POSSIBLE SECOND-GENERATION EFFECTS

If children are affected by their mothers' implants, they are exposed either in utero through transplacental passage or after birth through breast-feeding. A study by Schmiterlow and Sjogren (1975) using intravenous radiolabeled polydimethylsiloxane in mice demonstrated that the compound or its metabolites pass across the placental barrier. We have evaluated several placentas from women with silicone implants and were not able to demonstrate either silicone particles in the tissue or a consistent histologic picture that would suggest an ongoing inflammatory reaction. Although photomicrographs have demonstrated silicone-like material in milk ducts of breast tissue, no blinded, case-controlled study has shown silicone in breast milk. We have recently evaluated several samples of breast milk from women with silicone implants and controls and have not noted any significant differences in silicon levels. Long-term studies are needed to determine whether silicone can pass into the breast milk and to correlate silicone levels with the subsequent development of clinical disease in the children.

CLINICAL SPECTRUM OF DISEASE

Many studies have described clinical findings in women with breast implants. The spectrum of disease has ranged from classical scleroderma to nonspecific rheumatologic complaints including fibromyalgia and chronic fatigue syndrome (Spiera, 1988; Spiera & Kerr, 1993; Sahn et al., 1990; Vasey et al., 1994; Solomon, 1994; Bridges et al., 1993). In addition, several other syndromes have been described, including atypical chest pain syndrome (Cuellar et al., 1994; Silver et al., 1994), sicca syndrome (Freundlich et al., 1994; Nardella, 1995), and nonspecific neurologic disorders (Ostermeyer-Shoaib & Patten, 1995). However, other studies have not found an association between silicone implants and classic or atypical rheumatologic diseases (Sanchez-Guerrero et al., 1994; Strom et al., 1994; Gabriel et al., 1994; Duffy & Woods, 1994).

A large-scale questionnaire has been in use over the past few years in an effort to examine possible clinical diseases in children due to their mothers' implants. This questionnaire was designed to obtain information about affected children and their parents, to determine whether specific or unusual patterns of disease exist in this group of children, and to look for clinical signs or symptoms that can distinguish significantly ill children from those that are not as sick. Due to the selection bias in patient recruitment, disease incidence cannot be determined from the questionnaire data. Preliminary results comparing postimplant children ($n = 252$) with preimplant children ($n = 127$) from 190 women with breast implants are that the general clinical health in the postimplant children was significantly worse ($p < .001$). Analysis of specific symptoms in postimplant children has demonstrated that intestinal complaints are most common, followed by nonspecific symptoms such as fatigue and weakness and joint complaints.

AUTOIMMUNE DISEASE

Studies have suggested that silicone polymers may have adjuvant properties (Yoshida et al., 1994; Naim et al., 1993). Additionally, as silicon is an essential component of connective tissue (Carlisle, 1984), silicone-induced immune responses may preferentially affect connective tissue. Due to these factors, concern has been raised that silicone leaking from an implant may lead to the development of autoimmune disease (Goldblum et al., 1992; Kacew, 1994).

We recently demonstrated scleroderma-like esophageal dysmotility in children breast-fed by mothers with silicone breast implants (Levine & Ilowite,

1994a). In this preliminary study, manometric abnormalities were seen only in breast-fed children, whereas bottle-fed children had findings that were similar to controls. Although the number of patients studied was small, similar findings were noted in a larger cohort of patients (Levine & Ilowite, 1994b). Although none of the patients reported have progressed to classical scleroderma, concern has been raised that the pathophysiologic mechanisms regarding development of this disorder may be similar to the clinical disease in women and may involve an immunologic response to the silicone implant.

AUTOANTIBODY PRODUCTION

Several aspects of the immune system response have been investigated in women with silicone breast implants. Initial studies have focused on the B-cell response to silicone and the presence of autoantibodies. The results of autoantibody production in these women are highly variable (Goldblum et al., 1992; Varga et al., 1989; Spiera & Kerr, 1993; Bridges et al., 1993; Teuber et al., 1993). The autoantibodies demonstrated have included antibodies to nRNP (Bridges et al., 1993), centromere (Spiera & Kerr, 1993; Bridges et al., 1993), high-titer antinuclear antibodies (ANA) (Teuber et al., 1993; Press et al., 1992), anticollagen antibodies (Teuber et al., 1993), and antisilicone antibodies (Wolf et al., 1993). In addition, Kossovsky et al. (1993) demonstrated antibodies against novel silicone-associated antigens. We have evaluated the prevalence of autoantibodies in children born to mothers with implants and have not found increased levels compared to controls (Levine et al., 1996a). In addition, we were not able to demonstrate an association between autoantibody positivity and clinical symptoms, or any relationship between autoantibody prevalence and methods of exposure (i.e., breast-feeding versus bottle-fed children).

CELL-MEDIATED IMMUNITY

Increased cell-mediated reactivity by T lymphocytes against silicon dioxide has been found in women with silicone implants compared to controls (Ojo-Amaize et al., 1994; Smalley et al., 1995a, 1995b; Ciapetti et al., 1995). In addition, flow cytometric and cell depletion analysis has shown that the responding cells are $CD4^+$ cells (Ojo-Amaize et al., 1994; Granchi et al., 1995). These results are similar to T-cell immune responses to other light metals (Saltini et al., 1989). The involvement of $CD4^+$ cells suggests that autoantibody production may be via amplification of T-cell help for autoreactive B cells. There have not been any published studies regarding cellular immune response in children of women with implants. However, we have preliminary data suggesting that there is increased T-cell stimulation to silicon dioxide in these children compared to controls. Follow-up studies are needed to confirm these findings as well as to understand their significance, especially regarding the association with clinical symptoms.

SILICONE LEVELS

Teuber et al. (1994) demonstrated elevated serum silicon levels in women with silicone gel implants compared to control women. In this study, the serum was assayed by inductively coupled plasma atomic emission spectroscopy for elemental silicon. However, the silicon levels did not correlate with the history of implant rupture or number of years with the implants. There have not been any studies that have evaluated silicon levels in children born to mothers with implants. We have suggested that the previously reported scleroderma-like esophageal motility disorder may be due to the leakage of silicon into breast milk and the subsequent transfer via breast-feeding. Currently, we are in the process of analyzing silicon levels in these children to determine whether they have an increased silicon load compared to controls. In addition, we are seeking to determine whether the esophageal motility abnormality is related to silicone by correlating silicon levels with the esophageal motility findings.

MACROPHAGE ACTIVATION

Histopathological examination of tissue surrounding silicone breast implants reveals the presence of "foamy" macrophages containing silicone-like material, multinucleated giant cells, and granulomas (Silver et al., 1993; Bridges et al., 1993; Gayou, 1979). These findings suggest a link between macrophage activation and silicone-induced inflammation. Phagocytosis of foreign material by macrophages leads to activation and release of inflammatory mediators such as nitric oxide (Adams & Hamilton, 1984). We recently measured urinary nitrite excretion in children born to mothers with silicone implants and found significantly increased levels compared to controls (Levine et al., 1996b). In addition, we found significantly increased nitrite production in vitro using a macrophage cell line in the presence of silicone compared to cells grown without silicone exposure.

CONCLUSIONS

The possibility that there is a second-generation effect from maternal silicone breast implants has not been conclusively proven. In addition, if children are affected, the clinical manifestations are incompletely understood. There have not been any studies that have attempted to look at an incidence to the problem. Large-scale and well-designed clinical and epidemiologic studies are needed to answer these important concerns. However, several reports suggest that there may be an association between breast implants and clinical disease in the children. Immunologic abnormalities that may be active in women with implants may also exist in their offspring, and these may be associated with the clinical findings in affected children. Further studies currently underway may help explain the connection between silicone implants and active disease in children years after exposure.

REFERENCES

Adams, D. O., and Hamilton, T. A. 1984. The cell biology of macrophage activation. *Annu. Rev. Immunol.* 2:283–318.

Angell, M. 1992. Breast implants—Protection or paternalism? *N. Engl. J. Med.* 326:1695–1696.

Berlin, C. M. 1994. Silicone breast implants and breast-feeding. *Pediatrics* 94:547–549.

Bridges, A. J., and Vasey, F. B. 1993. Silicone breast implants: History, safety and potential complications. *Arch. Int. Med.* 153:2638–2644.

Bridges, A. J., Conley, C., Wang, G., Burns, D. E., and Vasey, F. B. 1993. A clinical and immunologic evaluation of women with silicone breast implants and symptoms of rheumatic disease. *Ann. Int. Med.* 118:929–936.

Carlisle, E. M. 1984. Silicon. In *Biochemistry of the essential ultratrace elements*, ed. E. Freiden, pp. 257–291. New York: Plenum Press.

Ciapetti, G., Granchi, D., Stea, S., Cenni, E., Schiavon, P., Giuliani, R., and Pizzoferrato, A. 1995. Assessment of viability and proliferation of in vivo silicone-primed lymphocytes after in vitro re-exposure to silicone. *J. Biomed. Mater. Res.* 29:583–590.

Cuellar, M. L., Garcia, C., Molina, J. F., and Espinoza, L. R. 1994. Angina-like chest pain in women with silicone breast implants. *Arthritis Rheum.* 37:S270.

Domanskis, E. J., and Owsley, J. Q. 1976. Histological investigation of the etiology of capsular contracture following augmentation mammoplasty. *Plastic Reconstr. Surg.* 58:689–693.

Duffy, M. J., and Woods, J. E. 1994. Health risks of failed silicone gel breast implants: A 30-year clinical experience. *Plastic Reconstr. Surg.* 94:295–299.

Dunn, K. W., Hall, P. N., and Khoo, C. T. 1992. Breast implant materials: Sense and safety. *Br. J. Plastic Surg.* 45:315–321.

Freundlich, B., Altman, C., Sandorfi, N., Greenberg, M., and Tomaszewski, J. 1994. A profile of symptomatic patients with silicone breast implants: A Sjogrens-like syndrome. *Semin. Arthritis Rheum.* 24:44–53.

Gabriel, S.E., O'Fallon, W. M., Kurland, L. T., Beard, C. M., Woods, J. E., and Melton, L. J. 1994. Risk of connective tissue diseases and other disorders after breast implantation. *N. Engl. J. Med.* 330:1697–1702.

Gayou, R. M. 1979. A histological comparison of contracted and non-contracted capsules around silicone breast implants. *Plastic Reconstr. Surg.* 63:700–703.

Goldblum, R. M., Pelley, R. P., O'Donell, A. A., Pyron, D., and Heggers, J. P. 1992. Antibodies to silicone elastomers and reactions to ventriculoperitoneal shunts. *Lancet* 340:510–513.

Granchi, D., Cavedagna, D., Ciapetti, G., Stea, S., Cenni, E., Schiavon, P., Giuliani, R., and Pizzoferrato, A. 1995. Silicone breast implants: The role of immune system on capsular contraction formation. *J. Biomed. Mater. Res.* 29:197–202.

Kacew, S. 1994. Current issues in lactation: Advantages environment, silicone. *Biomed. Environ. Sci.* 7:307–319.

Kessler, D. A. 1992. The basis of the FDA's decision on breast implants. *N. Engl. J. Med.* 326:1713–1715.

Kossovsky, N., Zeidler, M., Chun, G., Papasian, N., Nguyen, A., Rajguru, S., Stassi, J., Gelman, A., and Sponsler, E. 1993. Surface dependent antigens identified by high binding avidity of serum antibodies in a subpopulation of patients with breast prostheses. *J. Appl. Biomater.* 4:281–288.

LeVier, R. R., Harrison, M. C., Cook, R. R., and Lane, T. H. 1993. What is silicone? *Plastic Reconstr. Surg.* 92:163–167.

Levine, J. J., and Ilowite, N. T. 1994a. Scleroderma-like esophageal disease in children breast-fed from mothers with silicone breast implants. *J. Am. Med. Assoc.* 271:213–216.

Levine, J. J., and Ilowite, N. T. 1994b. Scleroderma-like esophageal disease in children of mothers with silicone breast implants. *J. Am. Med. Assoc.* 272:769–770.

Levine, J. J., Lin, H. C., Rowley, M., Cook, A., Teuber, S. S., and Ilowite, N. T. 1996a. Lack of autoantibody expression in children born to mothers with silicone breast implants. *Pediatrics* 97:243–245.

Levine, J. J., Ilowite, N. T., Pettei, M. J., and Trachtman, H. 1996b. Increased urinary NO_3^- and NO_2^- and neopterin excretion in children breast fed by mothers with silicone implants: Evidence for macrophage activation. *J. Rheumatol.* 23:1083–1087.

Naim, J. O., Lanzafame, R. J., and van Oss, C. J. 1993. The adjuvant effect of silicone-gel on antibody formation in rats. *Immunol. Invest.* 22:151–161.

Nardella, F. A. 1995. Oral and ocular sicca syndrome in women with silicone breast implants. *Arthritis Rheum.* 38:S264.

Ojo-Amaize, E. A., Conte, V., Lin, H. C., Brucker, R. F., Agopian, M. S., and Peter, J. B. 1994. Silicone-specific blood lymphocyte response in women with silicone breast implants. *Clin. Diagn. Lab. Immunol.* 1:689–695.

Ostermeyer-Shoaib, B., and Patten, B. M. 1995. Multiple sclerosis-like syndrome in women with silicone breast implants: A novel neurological disease with rheumatological symptoms. *Arthritis Rheum.* 38:S264.

Pfleiderer, B., Ackerman, J. L., and Garrido, L. 1993. Migration and biodegradation of free silicone from silicone gel-filled implants after long-term implantation. *Magnetic Resonance Med.* 30:534–543.

Press, R. I., Peebles, C. L., Kumagai, Y., Ochs, R. L., and Tan, E. M. 1992. Antinuclear autoantibodies in women with silicone breast implants. *Lancet* 340:1304–1307.

Robinson, O. G., Bradley, E. L., and Wilson, D. S. 1995. Analysis of explanted silicone implants: A report of 300 patients. *Ann. Plastic Surg.* 34:1–7.

Sahn, E. E., Garen, P. D., Silver, R. M., and Maize, J. C. 1990. Scleroderma following augmentation mammoplasty: Report of a case and review of the literature. *Arch. Dermatol.* 126:1198–1202.

Saltini, C., Winestock, K., Kirby, M., Pinkston, P., and Crystal, R. G. 1989. Maintenance of alveolitis in patients with chronic beryllium disease by beryllium-specific helper T cell. *N. Engl. J. Med.* 320:1103–1109.

Sanchez-Guerrero, J., Schur, P. H., Sergent, J. S., and Liang, M. H. 1994. Silicone breast implants and rheumatic disease: Clinical, immunologic and epidemiologic studies. *Arthritis Rheum.* 37:158–168.

Sanger, J. R., Matloub, H. S., Yousif, N. J., and Komorowski, R. 1992. Silicone gel infiltration of a peripheral nerve and constrictive neuropathy following rupture of a breast prosthesis. *Plastic Reconstr. Surg.* 89:949–952.

Schmiterlow, C. G., and Sjogren, C. 1975. The distribution of ^{14}C-labelled KABI 1774. *Acta Pharmacol. Toxicol.* 36:131–138.

Silver, R. M., Sahn, E. E., Allen, J. A., Sahn, S., Greene, W., Maize, J. C., and Garen, P. D. 1993. Demonstration of silicon in sites of connective-tissue disease in patients with silicone-gel breast implants. *Arch. Dermato.* 129:63–68.

Silver, D., Silverman, S. L., and Mendoza, M. 1994. Chest wall syndrome in patients with silicone breast implants. *Arthritis Rheum.* 37:S270.

Smalley, D. L., Shanklin, D. R., Hall, M. F., Stevens, M. V., and Hanissian, A. 1995a. Immunologic stimulation of T lymphocytes by silica after use of silicone mammary implants. *FASEB J.* 9:424–427.

Smalley, D. L., Shanklin, D. R., Hall, M. F., and Stevens, M. V. 1995b. Detection of lymphocyte stimulation with silicone dioxide. *Int. J. Occup. Med. Toxicol.* 4:63–69.

Solomon, G. 1994. A clinical and laboratory profile of symptomatic women with silicone breast implants. *Semin. Arthritis Rheum.* 24:29–37.

Spiera, H. 1988. Scleroderma after silicone augmentation mammoplasty. *J. Am. Med. Assoc.* 260:236–238.

Spiera, H., and Kerr, L. D. 1993. Scleroderma following silicone implantation: A cumulative experience of 11 cases. *J. Rheumatol.* 20:958–961.

Strom, B. L., Reidenberg, M. M., Freundlich, B., and Schinnar, R. 1994. Breast silicone implants and risk of systemic lupus erythematosus. *J. Clin. Epidemiol.* 47:1211–1214.

Teuber, S. S., Rowley, M. J., Yoshida, S. H., Ansari, A. A., and Gershwin, M. E. 1993. Anti-collagen autoantibodies are found in women with silicone breast implants. *J. Autoimmunol.* 6:367–377.

Teuber, S. S., Saunders, R. L., Gershwin, M. E., Halpern, G. M., Brucker, R. F., Conte, V., Goldman, B. D., Wood, W. G., and Winger, E. E. 1994. Elevated serum silicone levels in women with silicone gel breast implants. *Arthritis Rheum.* 37:S422.

Teuber, S. S., Ito, L. K., Anderson, M., and Gershwin, M. E. 1995. Silicone breast implant-associated scarring dystrophy of the arm. *Arch. Dermatol.* 131:54–56.

Varga, J., Schumacher, R., and Jimenez, S. A. 1989. Systemic sclerosis after augmentation mammoplasty with silicone implants. *Ann. Int. Med.* 111:377–383.

Vasey, F. B., Havice, D. L., Bocanegra, T. S., Seleznick, M. J., Bridgeford, P. H., Martinez-Osuna, P., and Espinoza, L. R. 1994. Clinical findings in symptomatic women with silicone breast implants. *Semin. Arthritis Rheum.* 24:22–28.

Wells, A. F., Daniels, S., Wells, K. E., and Vasey, F. B. 1994. IL-6 and multinucleated giant cell in capsular biopsies from patients with silicone breast implants. *Arthritis Rheum.* 37:S422.

Winding, O., Christensen, L., Thomsen, J. L., Nielsen, M., Breiting, V., and Brand, T. B. 1988. Si in human breast tissue surrounding silicone gel prostheses. *Scand. J. Plastic Surg.* 22:127–130.

Wolf, L. E., Lappe, M., Peterson, R. D., and Ezrailson, E. G. 1993. Human immune response to polydimethylsiloxane (silicone): Screening studies in a breast implant population. *FASEB J.* 7:1265–1268.

Yoshida, S. H., Teuber, S. S., German, J. B., and Gershwin, M. E. 1994. Immunotoxicity of silicone: Implications of oxidant balance towards adjuvant activity. *Food Chem. Toxicol.* 32:1089–1100.

Chapter 8

Aspartame and Development

Bennett A. Shaywitz

Aspartame (APM, L-aspartyl-L-phenylalanine methyl ester) is a dipeptide sweetener approximately 180–200 times sweeter than sucrose. Approximately 200 safety and metabolism studies have been conducted on APM and its degradants in animals, in vitro, and in humans. These studies have been summarized in various monographs and reviews (Stegink & Filer, 1984; Janssen & van der Heijden, 1988; Butchko & Kotsonis, 1989; Jobe & Dailey, 1993; Meldrum, 1993; Lajtha et al., 1994). These summaries have included assessments of APM administered during reproduction and development in animals (Stegink & Filer, 1984; Butchko & Kotsonis, 1989; Meldrum, 1993; Lajtha et al., 1994) and discussions of clinical studies done with APM during pregnancy and in infants and children (Stegink & Filer, 1984; Janssen & van der Heijden, 1988; London, 1988; Butchko & Kotsonis, 1989; Jobe & Dailey, 1993; Meldrum, 1993; Lajtha et al., 1994). No adverse consequences of APM administration have been found at doses hundreds to thousands times greater than those encountered after normal APM consumption. Metabolic studies have demonstrated that APM is metabolized in the gastrointestinal tract to its three components: aspartic acid (ASP), phenylalanine (PHE), and methanol (Stegink & Filer, 1984; Stegink et al., 1989). These components normally occur in the diet in much greater amounts than derived from APM in food. The metabolic disposition of APM has been demonstrated to be consistent in infants and adults (Filer et al., 1983).

LEVELS OF EXPOSURE TO APM IN DIET

A no-observed-effect level (NOEL) of greater than 2000 to 4000 mg/kg body weight was established for APM in animals after comprehensive assessment of the sweetener in toxicology and pharmacology studies. The safety and metabolism of APM, as well as the pharmacokinetics of its components, have been extensively studied in humans as well. The results of clinical studies indicate that APM exposures at doses approximately 50–100 times actual daily consumption levels (90th percentile, 14-day average consumptions) are not associated with adverse health effects in any age group.

Regulatory bodies typically assign an acceptable daily intake (ADI) for food additives prior to approval. The ADI is not an absolute limit but is the consumption level considered safe on a chronic basis; occasionally exceeding the ADI is not considered harmful (Tollefson & Barnard, 1992). The ADI (usually expressed in mg additive/kg body weight) is most often the NOEL from animal toxicology studies divided by a safety factor of 100. The Scientific Committee for Foods of the European Economic Community and the Joint Expert Committee for Food Additives of the Food and Agriculture Organization/ World Health Organization (JECFA) thus established an ADI for APM of 40 mg/kg/day. Subsequently, in 1983, the Food and Drug Administration (FDA) in the United States established an ADI for APM of 50 mg/kg/day based on additional data available from humans. Either of these ADIs represents a high level of consumption of APM that is impossible from a single bolus consumption from foods and is unlikely even over the course of a day (Butchko & Kotsonis, 1991; Butchko et al., 1994; Tollefson & Barnard, 1992).

METHANOL COMPONENT OF APM

APM is approximately 10% methanol by weight, and methanol is formed from the hydrolysis of the phenylalanine methyl ester moiety in the intestinal lumen. It has been speculated that such APM-derived methanol may be potentially neurotoxic to the brain and retina (Monte, 1984). However, the amounts of methanol released from APM are well below those of normal dietary exposures to methanol in fruits, vegetables, and juices (Lajtha et al., 1994; Stegink & Filer, 1984). Further, direct administrations of enormous doses of APM have not been shown to raise blood concentrations of formate, the toxic metabolite of methanol, in humans (Stegink et al., 1981, 1983, 1989; Shahangian et al., 1984). The sensitivity of conventional analytical techniques is inadequate to detect formed methanol in human adults or children after APM dosages up to 34 mg/kg, more than 10 times APM consumption levels at the 90th percentile. Thus, the methanol component from APM would have no adverse health consequences during development, during pregnancy, or in adults.

ASPARTIC ACID COMPONENT OF APM

Issues were raised regarding the ASP component of APM, particularly as an excitotoxicant in neonates (Olney, 1983). The ASP moiety is approximately 40% of APM by weight. ASP and the other dicarboxylic amino acid, glutamate (GLU), comprise about 20–25% of dietary protein. The level of aspartic acid in brain is much higher than in the blood; thus, a severalfold increase in blood ASP does not result in an increase in brain concentrations of this amino acid (Battistin et al., 1971). Large bolus doses of GLU or ASP administered orally are capable of causing excitatory damage to the arcuate nucleus in neonatal rodents, although this has not been demonstrated for either of these amino acids when they are administered to rodents with food (Meldrum, 1993). Whether a similar excitotoxic mechanism is operative in primates following bolus administration of GLU or ASP is unlikely. Olney and co-workers (Olney & Sharpe, 1969; Olney et al., 1972) reported this phenomenon in neonatal nonhuman primates administered large bolus doses of glutamic acid. This observation has not been reproduced by a number of other scientists with either glutamic acid or aspartame at higher dosages (Newman et al., 1973; Wen et al., 1973; Goldberg et al., 1974; Abraham et al., 1975; Reynolds et al., 1979, 1980, 1984).

The human infant is apparently well adapted to a high intake of free glutamate; mature human milk is unique among mammalian milks in its high content of free amino acids. Glutamate and taurine are the free amino acids in highest concentrations in human milk and are in greater concentration than in other mammalian species including primates (Harzer et al., 1984; Pamblanco et al., 1989). The 10-fold higher concentration of free excitatory amino acids in human milk than in maternal plasma is apparently independent of excitatory amino acids being consumed in the diet of the lactating female (Baker et al., 1979; Meldrum, 1993). In utero, the human placenta acts as a highly efficient barrier to fetal exposure to excitatory amino acids (Pitkin et al., 1979).

Excitotoxic damage after bolus administration of GLU and/or ASP in neonatal rodents occurs only when plasma ASP concentrations exceed 1100 µmol/ L, or plasma GLU concentrations exceed 750 µmol/L, or the combined plasma concentrations of ASP plus GLU exceed 1280 µmol/L (Stegink & Filer, 1984; Meldrum, 1993). In the human adult, fasting plasma concentrations are around 20–25 µmol/L for GLU and are 1–5 µmol/L for ASP (Perry et al., 1975, 1990; Stegink et al., 1986; Plaitakis & Caroscio, 1987). Plasma concentrations of both ASP and GLU in humans remain within postprandial ranges even after large doses of APM. High-dose APM (34 mg/kg) produced no change in fasting plasma ASP levels in healthy adults (Stegink & Filer, 1984). Acute bolus dosing of APM in healthy adults (up to 200 mg/kg), 1-year-old infants (up to 100 mg/kg), and phenlyketonuric (PKU) heterozygotes (up to 100 mg/kg) produced no significant changes in plasma ASP concentrations when APM doses were less than 100 mg/kg. Small increases in plasma ASP concentrations occurred when APM doses were 100 mg/kg or greater but remained within normal postprandial ranges. Chronic administration of high doses of APM (75 mg/kg/day for 24 weeks) also did not change mean fasting plasma ASP concentrations (Leon et al., 1989). Thus, APM consumption even at very high levels does not present concern for adverse effects from excitotoxic mechanisms in developing infants, during pregnancy, or in adults (Meldrum, 1993; Lajtha et al., 1994).

PHENYLALANINE COMPONENT OF APM

Phenylalanine (PHE) is an essential amino acid comprising approximately 5% of dietary protein and makes up approximately 50% of APM by weight. PHE either is utilized in protein synthesis or is converted in the liver to tyrosine by the enzyme PHE hydroxylase. Tyrosine, in turn, is utilized in protein synthesis as well as being the precursor of catecholamine neurotransmitters in the central nervous system and norepinephrine and epinephrine in the adrenal medulla. At reported consumption levels (Butchko & Kotsonis, 1991), the PHE component of APM provides only approximately 1% of a 4-year-old child's daily dietary intake of PHE and approximately 2% of an adult's. Acute, repeated, and long-term studies have been done in humans to evaluate the safety of the PHE component of APM. After a 34-mg/kg bolus dose of APM in adults, peak plasma PHE concentrations were approximately 120 µmol/L but remained within normal postprandial range (90–150 µmol/L). After 200 mg/kg of APM given to healthy adults, plasma PHE concentrations achieved a peak of approximately 490 µmol/L and declined toward baseline over several hours (Steglink & Filer, 1984). In humans, APM administered as bolus or as repeated doses to normal adults, phenylketonuric (PKU) heterozygotes, 1-year-old infants, or insulin-dependent or non-insulin-dependent diabetics does not increase plasma PHE levels above those experienced after ingesting a protein-containing meal (Filer & Stegink, 1989; Gupta et al., 1989).

PHE is one of a group of large neutral amino acids (LNAA) that compete for entry into the brain through a carrier transport system. The rate of entry of PHE into the brain is predicted by the ratio of the plasma concentration of PHE to the sum of the plasma concentrations of the other LNAA (Maher & Wurtman, 1987; Lajtha et al., 1994). Fluctuations in the ratio of PHE to other LNAA occur as the result of dietary manipulations and normal variability in diet (Fernstrom et al., 1979). Since APM is a source of PHE without other competing amino acids, it has been suggested that APM may lead to increases in the PHE/LNAA ratio, and thus, in brain PHE concentrations. It has been further speculated that such changes in brain PHE concentrations would result in changes in brain neurotransmitters, seizures, behavioral changes, or cognitive function (Maher & Wurtman, 1987; Pardridge, 1986). While isolated effects of APM on brain neurotransmitter or metabolite concentrations have been noted in a few animal studies, careful review of the total literature demonstrates these effects are neither consistent nor reproducible (Lajtha et al., 1994). More definitive studies evaluating the receptor kinetics or release of brain neurotransmitters, considered more reliable indications of neurotransmission than static steady-state concentrations, have failed to show a treatment effect of large doses of APM or PHE (Lajtha et al., 1994). Similarly, animal studies evaluating seizure susceptibility, behavior, and central nervous system development have shown no reproducible effects following acute or chronic administrations of high doses of APM (Stegink & Filer, 1984; Butchko & Kotsonis, 1989; Jobe & Dailey,

1993). The absence of findings following APM administration are consistent with the fact that plasma amino acid concentrations and ratios fluctuate as a consequence of variations in the protein and carbohydrate content of meals. The brain is often exposed to fluctuations in the ratios of LNAAs under physiological conditions and is thus unlikely to respond to even abuse levels of APM consumption with important central effects (Lajtha et al., 1994).

DEVELOPMENTAL STUDIES WITH ASPARTAME IN ANIMALS

The effects of APM on morphological and behavioral development have been evaluated in both rodents and nonhuman primates (Brunner et al., 1979; Potts et al., 1980; Butcher & Vorhees, 1984; Reynolds et al., 1984; Suomi, 1984; Mahalik & Gautieri, 1984; Tilson et al., 1988; Holder, 1989; McAnulty et al., 1989). No neurobehavioral or developmental deficits were observed at dosages of APM less than 6000 mg/kg body weight (Brunner et al., 1979; Potts et al., 1980). In two-generation reproduction, perinatal, and postnatal toxicity studies in rodents, there were no delays in the onset of developmental milestones at dietary dosages of up to 4000 mg/kg body weight, the highest dosage administered to dams (Molinary, 1984).

Holder (1989) administered APM in drinking water to rats from conception to 38 days of age. The highest concentration provided a dosage of approximately 3500 mg/kg/day and produced no effect on morphological and reflex development, and spatial memory. Butcher and Vorhees (1984) examined the effects of APM on behavioral development in rats at prenatal doses of 1500–5000 mg/kg/day to dams and dosages of approximately 3000–9000 mg/kg postnatally to pups. Evaluation of APM in an extensive test battery found that only the dosage of APM approximating 9000 mg/kg was capable of producing adverse effects, and those were indistinguishable from equimolar amounts of L-PHE. Cross-fostering experiments demonstrated that even at the highest possible administrable dosages of PHE, no adverse effects occurred during prenatal exposure (Brunner et al., 1979); this is probably due to the dams' high capacity to hydroxylate PHE (Vorhees et al., 1981; Hjelle et al., 1992).

Oral administration of APM to pregnant mice at dosages up to 4000 mg/kg had no adverse effects either to dams or offspring. Developmental and reflex ontogeny proceeded at a normal rate in pups exposed to APM in utero. Specific evaluations of visual and other sensory systems indicated that these systems were not altered after in utero exposure of mice to APM (McAnulty et al., 1989). This work was an attempt to replicate an earlier study that had reported an apparent delay in developing the visual placing response after neonatal mice were exposed to aspartame in utero (Mahalik & Gautieri, 1984). The fact that the study reporting the presumable developmental delay apparently did not include concurrent controls nor uniform litter sizes may explain the different results obtained.

Infant primates administered APM dosages up to 3000 mg/kg body weight in diet showed no marked changes in growth rate, developmental milestones, learning, or sensory discrimination (Reynolds et al., 1984; Suomi, 1984). Reynolds and co-workers (1984) dosed stump-tail macaques for the first 9 months of postnatal development with 1000, 2000, or 3000 mg/kg of APM in diet. Growth and maturational indices were essentially identical for all infant monkeys regardless of APM dosage, and there were no APM-related effects. At the conclusion of dosing, the young macaques were transferred to the Wisconsin Primate Laboratory to be evaluated by Suomi (1984) for long-term behavioral effects following developmental dosing with APM. The behavioral test battery included object discrimination, pattern discrimination, and learning and hearing assessments. No evidence of changes or deficits in any monkey having had early dietary exposure to APM was found.

ASPARTAME CONSUMPTION BY PREGNANT FEMALES

The results of animal reproductive studies, the numerous tolerance studies in humans, the extensive human pharmacokinetic data available for APM, the metabolism of APM to usual dietary constituents, and the widespread consumption of APM by pregnant women without adverse effect all indicate that APM is safe during pregnancy. Initial hypothetical concerns regarding APM consumption during pregnancy were due largely to concerns regarding the PHE moiety in APM. A great deal of information is available regarding safe plasma concentrations of PHE during pregnancy for individuals homozygous for PKU (frank deficiency in PHE metabolism) or those individuals heterozygous for PKU (somewhat compromised PHE metabolism). Normal individuals can never achieve plasma concentrations of PHE dangerous to themselves or the fetus by dietary manipulation alone (London, 1988; Butchko & Kotsonis, 1989; Lajtha et al., 1994).

Offspring of homozygous PKU mothers who are not on a PHE-restricted diet will usually suffer adverse consequences from the mother's elevated plasma PHE concentrations during pregnancy. For these individuals, rigid dietary restriction of PHE needs to be instituted prior to conception (Drogari et al., 1987), and APM should be factored into this PHE-restricted diet. Conversely, in pregnant women homozygous for PKU but not on a PHE-restricted diet, in practical terms the already elevated levels of PHE are not likely to be further elevated by APM consumption. Cabellero and co-workers (1986) found in untreated PKU homozygotes and mild hyperphenylalaninemics given 10 mg/kg of APM, probably the largest dose of APM a person would be exposed to at any one time, that there was no significant elevation beyond the already high baseline PHE values. Thus, when there is a risk from dietary PHE, the normal diet itself poses as great or greater risk to the fetus than does APM in foods. It is not possible for healthy individuals or individuals heterozygous for PKU to consume enough APM in foods to achieve, much less sustain, the plasma PHE concentrations associated with adverse outcomes in individuals homozygous for PKU (Stegink et al., 1977, 1979, 1989, 1990).

STUDIES WITH ASPARTAME IN INFANTS, CHILDREN, AND ADOLESCENTS

Tolerance studies in 1-year-old children given bolus doses of APM of 34, 50, or 100 mg/kg body weight demonstrated no adverse effect in infants and further showed that children metabolize APM as well as adults (Filer et al., 1983). Frey (1976) conducted a 13-week duration, double-blind, placebo-controlled study in 126 healthy children and adolescents ranging from 2 to 21 years of age. These children were given foods containing either APM or sucrose and, when necessary to achieve dosing objectives, additional capsules containing APM or sucrose. The doses approximated 28–77 mg/kg/day of APM over the 13-week dosing interval. There were no clinically significant differences between the APM and placebo groups in any of the clinical or biochemical parameters measured, including plasma PHE and tyrosine concentrations. Neither blood methanol nor urinary phenylpyruvic acid was detected during the study. Physical examinations, ophthalmologic examinations, and adverse experiences were similar whether receiving APM or placebo.

In another randomized, double-blind, placebo-controlled study with crossover design, we examined the effects of APM consumption on the cognitive and behavioral status of 15 children with a history of attention deficit disorder (ADD) (Shaywitz et al., 1994b). The 4-week study consisted of two 2-week periods that were identical except for administration of either APM (34 mg/kg/day) or placebo. APM at greater than 10 times usual consumption levels had no effect on behavior, cognitive function, adverse experiences, or urinary excretion of monoamines and metabolites compared to the placebo. The expected increase in plasma PHE and tyrosine did occur following APM administration (Shaywitz et al., 1994b).

In another randomized, double-blind, placebo-controlled study with crossover design, we examined epileptic children consuming APM or placebo over similar 2-week periods (Shaywitz et al., 1994a). The 10 children had generalized seizures (7 subjects), absence seizures (5 subjects including 4 with generalized seizures), or complex partial seizures (2 subjects). There were no significant differences between APM and placebo in any indications of seizure activity including standard electroencephalogram (EEG) and 24-h EEG monitoring. No differences were noted for any biochemical measures except for the expected increases in PHE and tyrosine after APM. Our findings indicate that in this group of vulnerable children, APM does not provoke seizures. In another randomized, double-blind, placebo-controlled study with crossover design that we conducted with 18 individuals who had seizures allegedly related to APM consumption (16 adults and 2 children), no clinical or electrical seizures occurred after ingestion of 50 mg/kg APM (Rowan et al., 1995). These results suggest that APM at this high acute dosage is no more likely than placebo to cause seizures in individuals who reported that their seizures were provoked by aspartame consumption.

ANECDOTAL REPORTS FOLLOWING
APM CONSUMPTION

The widespread use of a food additive such as APM makes it inevitable that adverse health events will occur coincidentally with the consumption of products containing that food additive (Tollefson & Barnard, 1992). Anecdotal reports of adverse experiences following APM consumption have been monitored and reviewed by epidemiologists at the Centers for Disease Control (CDC) and the U.S. FDA. These investigators concluded that (1) the complaints are generally mild and common in the general population, (2) there is no consistent pattern of symptoms that can causally be related to APM, and (3) only through focused clinical studies can these allegations be thoroughly studied (Butchko et al., 1994; Tollefson & Barnard, 1992). APM tolerance studies have been done in healthy adults, children, and adolescents, individuals heterozygous for PKU, and individuals who may be heavy users of APM, such as obese individuals and diabetics. In addition, studies have been done to evaluate the tolerance of APM by individuals who have altered amino acid and protein metabolism (e.g., individuals with liver and renal disease). The results of these studies have demonstrated that even large doses of APM are not associated with adverse effects either in healthy adults and children or in various subpopulations (Steginck & Filer, 1984; Janssen & van der Heijden, 1988; Butchko & Kotsonis, 1989; Gupta et al., 1989; Leon et al., 1989; Hertelendy et al., 1993).

When investigating allegations of adverse effects, clinical studies should be conducted under double-blind, placebo-controlled conditions. The results of double-blind, placebo-controlled studies show that in normal healthy humans, there is no alteration in cognition, mood, behavior, or electrophysiology that can be attributed to consumption of APM, even when doses are sufficient to raise the plasma PHE/LNAA ratio. APM has been shown to have no effects on behavior or cognition in pilots (Stokes et al., 1994) or in normal or hyperactive children, although some of these studies were designed to evaluate the effect of sugar on behavior and APM was used as the placebo (Wolraich et al., 1994). Studies have been conducted specifically in potentially vulnerable populations such as in individuals with headache (Koehler & Glaros, 1988; Schiffman et al., 1987; van den Eeden et al., 1994), seizures (Camfield et al., 1992; Shaywitz et al., 1994a; Rowan et al., 1995), affective disorders (Walton et al., 1993), attention deficit disorder (Shaywitz et al., 1994b; Wolraich et al., 1994), and individuals heterozygous for PKU (Trefz et al., 1994). In general, these studies in potentially vulnerable populations support the absence of effects following APM administration. One study suggested that ingestion of APM by migraineurs causes an increase in headache frequency for some subjects (Koehler & Glaros,

1988) although in a population of individuals reporting headaches after consuming products containing APM, in a controlled double-blind setting APM was no more likely to produce headache than was placebo (Schiffman et al., 1987). Two studies in potentially vulnerable populations provided suggestive results, one concerning the effects of APM on adults with depression (Walton et al., 1993) and the other reporting neurophysiological alterations in the EEGs of children with absence epilepsy (Camfield et al., 1992). In both studies, however, interpretation of the suggestive results is confounded on the bases of statistical design limitations and inadequately controlled experimental paradigms.

CONCLUSIONS

In the large majority of studies that have evaluated normal individuals, adults and children with seizures (including absence seizures), children with ADD or children considered sugar-reactive, and PKU heterozygous adults, there was no evidence of adverse effects from APM administration. Furthermore, animal studies done during gestation and development have failed to provide any experimental basis for neurotoxicological or developmental consequences from APM consumption. APM is chemically unique compared to other high-intensity sweeteners in that it is broken down in the intestinal lumen into components that are normal constituents of the diet. The extensive scientific literature on APM indicates that it can be safely consumed with no apparent consequences to fetal or postnatal development. Further, the results of animal reproductive studies, the numerous tolerance studies available in humans of all age groups, the extensive human pharmacokinetic data available for APM, and the long history of consumption of APM by pregnant women without adverse effects to either mother or child indicate that APM may be safely consumed during pregnancy.

REFERENCES

Abraham, R., Swart, J., Goldberg, L., and Coulston, F. 1975. Electron microscopic observations of hypothalami in neonatal rhesus monkeys (*Macaca mulatta*) after administration of monosodium L-glutamate. *Exp. Mol. Pathol.* 23:203–213.

Baker, G. L., Filer, L. J., and Stegink, L. D. 1979. Factors influencing dicarboxylic amino acid content of human milk. In *Glutamic acid: Advances in biochemistry and physiology*, eds. L. J. Filer, Jr., S. Garattini, M. R. Kare, W. A. Reynolds, and R. J. Wurtman, pp. 111–123. New York: Raven Press.

Battistin, L., Grynbaum, A., and Lajtha, A. 1971. The uptake of various amino acids by the mouse brain in vivo. *Brain Res.* 29:85–99.

Brunner, R. L., Vorhees, C. V., Kenney, L., and Butcher, R. E. 1979. Aspartame: Assessment of developmental psychotoxicity of a new artificial sweetener. *Neurobehav. Toxicol.* 1:79–86.

Butcher, R. E., and Vorhees, C. V. 1984. Behavioral testing in rodents given food additives. In *Aspartame: Physiology and biochemistry*, eds. L. D. Stegink and L. J. Filer, pp. 379–404. New York: Marcel Dekker.

Butchko, H. H., and Kotsonis, F. N. 1989. Aspartame: Review of recent research. *Comments Toxicol.* 3:253–278.

Butchko, H. H., and Kotsonis, F. N. 1991. Acceptable daily intake vs. actual intake: The aspartame example. *J. Am. Coll. Nutr.* 10:258–266.

Butchko, H. H., Tschanz, C., and Kotsonis, F. N. 1994. Postmarketing surveillance of food additives. *Regul. Toxicol. Pharmacol.* 20:105–118.

Caballero, B., Mahon, B. E., Rohr, F. J., Levy, H. L., and Wurtman, R. J. 1986. Plasma amino acid levels after single–dose aspartame consumption in phenylketonuria, mild hyperphenylalaninemia and heterozygous state for phenylketonuria. *J. Pediatr.* 109:668–671.

Camfield, P. R., Camfield, C. S., Dooley, J. M., Gordon, K., Jolleymore, S., and Weaver D. F. 1992. Aspartame exacerbates EEG spike-wave discharge in children with generalized absence epilepsy: A double-blind controlled study. *Neurology* 42:1000–1003.

Drogari, E., Beasley, M., Smith, I., and Lloyd, J. K. 1987. Timing of strict diet in relation to fetal damage in maternal phenylketonuria. *Lancet* 2:927–930.

Fernstrom, J. D., Wurtman, R. J., Hammarstrom-Wiklund, B., Rand, W. M., Munro, H. N., and Davidson, C. S. 1979. Diurnal variations in plasma concentrations of tryptophan, tyrosine, and other neutral amino acids: Effect of dietary protein intake. *Am. J. Clin. Nutr.* 32:1912–1922.

Filer, L. J., Jr., and Stegink, L. D. 1989. Aspartame metabolism in normal adults, phenylketonuric heterozygotes and diabetic subjects. *Diabetes Care* 12:67–74.

Filer, L. J., Jr., Baker, G. L., and Stegink, L. D. 1983. Effect of aspartame loading on plasma and erythrocyte free amino acid concentrations in one-year-old infants. *J. Nutr.* 113:1591–1599.

Frey, G. H. 1976. Use of aspartame by apparently healthy children and adolescents. *J. Toxicol. Environ. Health* 2:401–415.

Goldberg, L., Abraham, R., and Coulston, F. 1974. When is glutamate neurotoxic? *N. Engl. J. Med.* 290:1326–1327.

Gupta, V., Cochran, C., Parker, T. F., Long, D. L., Ashby, J., Gorman, M. A., and Liepa, G. U. 1989. Effect of aspartame on plasma amino acid profiles of diabetic patients with chronic renal failure. *Am. J. Clin. Nutr.* 49:1302–1306.

Harzer, G., Franzke, V., and Bindels, J. G. 1984. Human milk nonprotein nitrogen components: Changing patterns of free amino acids and urea in the course of early lactation. *Am. J. Clin. Nutr.* 40:303–309.

Hertelendy, Z. I., Mendenhall, C. L., Rouster, S. D., Marshall, L., and Weesner, R. 1993. Biochemical and clinical effects of aspartame in patients with chronic, stable alcoholic liver disease. *Am. J. Gastroenterol.* 88:737–743.

Hjelle, J. J., Dudley, R. E., Marietta, M. P., Sanders, P. G., Dickie, B. C., Brisson, J., and Kotsonis, F. N. 1992. Plasma concentrations and pharmacokinetics of phenylalanine in rats and mice administered aspartame. *Pharmacology* 44:48–60.

Holder, M. D. 1989. Effects of perinatal exposure to aspartame on rat pups. *Neurotoxicol. Teratol.* 11:1–6.

Janssen, P. J. C. M., and van der Heijden, C. A. 1988. Aspartame: Review of recent experimental and observational data. *Toxicology* 50:1–26.

Jobe, P. C., and Dailey, J. W. 1993. Aspartame and seizures. *Amino Acids* 4:197–235.

Koehler, S. M., and Glaros, A. 1988. The effect of aspartame on migraine headache. *Headache* 28:10–13.

Lajtha, A., Reilly, M. A., and Dunlop, D. S. 1994. Aspartame consumption: Lack of effects on neural function. *J. Nutr. Biochem.* 5:266–283.

Leon, A. S., Hunninghake, D. B., Bell, C., Rassin, D. K., and Tephly, T. R. 1989. Safety of long-term large doses of aspartame. *Arch. Intern. Med.* 149:2318–2324.

London, R. S. 1988. Saccharin and aspartame: Are they safe to consume during pregnancy? *J. Reproduct. Med.* 33:17–21.

Mahalik, M. P., and Gautieri, R. F. 1984. Reflex responsiveness of CF-1 mouse neonates following maternal aspartame exposure. *Res. Commun. Psychol. Psychiatry Behav.* 9:385–403.

Maher, T. J., and Wurtman, R. J. 1987. Possible neurologic effects of aspartame, a widely used food additive. *Environ. Health Perspect.* 75:53–57.

McAnulty, P. A., Collier, M. J., Enticott, J., Tesh, J. M., Mayhew, D. A., Comer, C. P., Hjelle, J. J., and Kotsonis, F. N. 1989. Absence of developmental effects in CF-1 mice exposed to aspartame in utero. *Fundam. Appl. Toxicol.* 13:296–302.

Meldrum, B. 1993. Amino acids as dietary excitotoxins: A contribution to understanding neuro-degenerative disorders. *Brain Res. Rev.* 18:293–314.

Molinary, S. V. 1984. Preclinical studies of aspartame in nonprimate animals. In *Aspartame: Physiology and biochemistry,* eds. L. D. Stegink and L. J. Filer, pp. 289–306. New York: Marcel Dekker.

Monte, W. C. 1984. Aspartame: Methanol and the public health. *J. Appl. Nutr.* 36:42–54.

Newman, A. J., Heywood, R., Palmer, A. K., Barry, D. H., Edwards, F. P., and Worden A. N. 1973. The administration of monosodium L-glutamate to neonatal and pregnant rhesus monkeys. *Toxicology* 1:197–204.

Olney, J. W. 1983. Excitotoxins: An overview. In *Excitotoxins,* eds. K. Fuxe, P. Roberts, and R. Schwarcz, pp. 82–96. London: Macmillan.

Olney, J. W., and Sharpe, L. G. 1969. Brain lesions in an infant rhesus monkey treated with monosodium glutamate. *Science* 166:386–388.

Olney, J. W., Sharpe, L. G., and Feigin, R. D. 1972. Glutamate-induced brain damage in infant primates. *J. Neuropathol. Exp. Neurol.* 31:464–488.

Pamblanco, M., Portoles, M., Paredes, C., Ten, A., and Comin, J. 1989. Free amino acids in preterm and term milk from mothers delivering appropriate- or small-for-gestational-age infants. *Am. J. Clin. Nutr.* 50:778–781.

Pardridge, W. M. 1986. Potential effects of the dipeptide sweetener aspartame on the brain. In *Nutrition and the brain,* vol. 7, eds. R. J. Wurtman and J. J. Wurtman, pp. 199–241. New York: Raven Press.

Perry, T. L., Hansen, S., and Kennedy, J. 1975. CSF amino acids and plasma-CSF amino acid ratios in adults. *J. Neurochem.* 24:587–589.

Perry, T. L., Krieger, C., Hansen, S., and Eisen, A. 1990. Amyotrophic lateral sclerosis: Amino acid levels in plasma and cerebrospinal fluid. *Ann. Neurol.* 28:12–17.

Pitkin, R. M., Reynolds, W. A., Stegink, L. D., and Filer, L. J. 1979. Glutamate metabolism and placental transfer in pregnancy. In *Glutamic acid: Advances in biochemistry and physiology,* eds. L. J. Filer, Jr., S. Garattini, M. R. Kare, W. A. Reynolds, and R. J. Wurtman, pp. 103–110. New York: Raven Press.

Plaitakis, A., and Caroscio, J. T. 1987. Abnormal glutamate metabolism in amyotrophic lateral sclerosis. *Ann. Neurol.* 22:575–579.

Potts, W. J., Bloss, J. L., and Nutting, E. F. 1980. Biological properties of aspartame. I. Evaluation of central nervous system effects. *J. Environ. Pathol. Toxicol.* 3:341–353.

Reynolds, W. A., Lemkey-Johnston, N., and Stegink, L. D. 1979. Morphology of the fetal monkey hypothalamus after in utero exposure to monosodium glutamate. In *Glutamic Acid: Advances in biochemistry and physiology,* eds. L. J. Filer, Jr., S. Garattini, M. R. Kare, W. A. Reynolds, and R. J. Wurtman, pp. 217–229. New York: Raven Press.

Reynolds, W. A., Stegink, L. D., Filer, L. J., Jr., and Renn, E. 1980. Aspartame administration to the infant monkey: Hypothalamic morphology and plasma amino acid levels. *Anat. Rec.* 198:73–85.

Reynolds, W. A., Bauman, A. F., Stegink, L. D., and Filer, L. J. 1984. Developmental assessment of infant macaques receiving dietary aspartame or phenylalanine. In *Aspartame: Physiology and biochemistry,* eds. L. D. Stegink and L. J. Filer, pp. 405–423. New York: Marcel Dekker.

Rowan, A. J., Shaywitz, B. A., Tuchman, L., French, J. A., Luciano, D., and Sullivan, C. M. 1995. Aspartame and seizure susceptibility: Results of a clinical study in reportedly sensitive individuals. *Epilepsia* 36:270–275.

Schiffman, S. S., Buckley, C. E., Sampson, H. A., Massey, E. W., Baraniuk, J. N., Follett, J. V., and Warwick, Z. S. 1987. Aspartame and susceptibility to headache. *N. Engl. J. Med.* 317:1181–1185.

Shahangian, S., Ash, K. O., and Rollins, D. E. 1984. Aspartame not a source of formate toxicity. *Clin. Chem.* 30:1264–1265.

Shaywitz, B. A., Anderson, G. M., Novotny, E. J., Ebersole, J. S., Sullivan, C. M., and Gillespie, S. M. 1994a. Aspartame has no effect on seizures or epileptiform discharges in epileptic children. *Ann. Neurol.* 35:98–103.

Shaywitz, B. A., Sullivan, C. M., Anderson, G. M., Gillespie, S. M., Sullivan, B., and Shaywitz, S. E. 1994b. Aspartame, behavior, and cognitive function in children with attention deficit disorder. *Pediatrics* 93:70–75.

Stegink, L. D., and Filer, L. J., Jr. 1984. *Aspartame: Physiology and biochemistry.* New York: Marcel Dekker.

Stegink, L. D., Filer L. J., Jr., and Baker, G. L. 1977. Effect of aspartame and aspartate loading upon plasma and erythrocyte free amino acid levels in normal adult volunteers. *J. Nutr.* 107:1837–1845.

Stegink, L. D., Filer L. J., Jr., Baker, G. L., and McDonnell, J. E. 1979. Effect of aspartame loading upon plasma and erythrocyte amino acid levels in phenylketonuric heterozygotes and normal adult subjects. *J. Nutr.* 109:708–717.

Stegink, L. D., Brummel, M. C., McMartin, K., Martin-Amat, G., Filer, L. J., Jr., Baker, G. L., and Tephly, T. R. 1981. Blood methanol concentrations in normal adult subjects administered abuse doses of aspartame. *J. Toxicol. Environ. Health* 7:281–290.

Stegink, L. D., Brummel, M. C., Filer, L. J., Jr., and Baker, G. L. 1983. Blood methanol concentrations in one-year-old infants administered graded doses of aspartame. *J. Nutr.* 113:1600–1606.

Stegink, L. D., Filer, L. J., Jr., Baker, G. L., and Bell, E. F. 1986. Plasma glutamate concentrations in 1-year-old infants and adults ingesting monosodium L-glutamate in consomme. *Pediatr. Res.* 20:53–58.

Stegink, L. D., Filer, L. J., Jr., Bell, E. F., Ziegler, E. E., and Tephly, T. R. 1989. Effect of repeated ingestion of aspartame-sweetened beverage on plasma amino acid, blood methanol, and blood formate concentrations in normal adults. *Metabolism* 38:357–363.

Stegink, L. D., Filer, L. J., Jr., Bell, E. F., Ziegler, E. E., Tephly, T. R., and Krause, W. L. 1990. Repeated ingestion of aspartame-sweetened beverages: Further observations in individuals heterozygous for phenylketonuria. *Metabolism* 39:1076–1081.

Stokes, A. F., Belger, A., Banich, M.T., and Bernadine, E. 1994. Effects of alcohol and chronic aspartame ingestion upon performance in aviation relevant cognitive tasks. *Aviat. Space Environ. Med.* 65:7–15.

Suomi, S. J. 1984. Effects of aspartame on the learning test performance of young stump-tail macaques. In *Aspartame: Physiology and biochemistry*, eds. L. D. Stegink and L. J. Filer, pp. 425–445. New York: Marcel Dekker.

Tilson, H. A., Zhao, D., Peterson, N. J., Nanry, K., and Hong, J. S. 1988. Behavioral and neurological effects of aspartame. In *Dietary phenylalanine and brain function*, eds. R. J. Wurtman and E. Ritter-Walker, pp. 104–110. Boston: Birkehauser.

Tollefson, L., and Barnard, R. J. 1992. An analysis of FDA passive surveillance reports of seizures associated with consumption of aspartame. *J. Am. Diet. Assoc.* 92:598–601.

Trefz, F., de Sonneville, L., Matthis, P., Benninger, C., Lanz-Englert, B., and Bickel, H. 1994. Neuropsychological and biochemical investigations in heterozygotes for phenylketonuria during ingestion of high dose aspartame (a sweetener containing phenylalanine). *Hum. Genet.* 93:369–374.

van den Eeden, S. K., Koepsell, T. D., Longstreth, W. T., Jr., van Belle, G., Daling, J. R., and McKnight, B. 1994. Aspartame ingestion and headaches: A randomized crossover trial. *Neurology* 44:1787–1793.

Vorhees, C. V., Butcher, R. E., and Berry, H. K. 1981. Progress in experimental phenylketonuria: A critical review. *Neurosci. Biobehav. Rev.* 5:177–190.

Walton, R. G., Hudak, R., and Green-Waite, R. J. 1993. Adverse reactions to aspartame: Double-blind challenge in patients from a vulnerable population. *Biol. Psychiatry* 34:13–17.

Wen, C., Hayes, K. C., and Gershoff, S. N. 1973. Effects of dietary supplementation of monosodium glutamate on infant monkeys, weanling rats and suckling mice. *Am. J. Clin. Nutr.* 26:803–813.

Wolraich, M. L., Lindgren, S. D., Stumbo, P. J., Steginck, L. D., Appelbaum, M. I., and Kiritsy, M. C. 1994. Effects of diets high in sucrose or aspartame on the behavior and cognitive performance of children. *N. Engl. J. Med.* 330:301–307.

Chapter 9

Over-the-Counter Medication Use and Toxicity in Children

Michael B. H. Smith and Michael D. Kogan

Minor infections and illness are a common family experience and are often treated with over-the-counter (OTC) medication. A visit to any local drugstore will reveal literally hundreds of products for various ailments. The presence of these packaged substances in a pharmacy often seems to imply a scientific effectiveness and proven safety record. However, many of these products have been poorly evaluated and are often not recommended by health care professionals. Many parents have misconceptions about their use and inadvertently use them in inappropriate ways. Side effects may occur even in the recommended doses or if used for too long or in young infants. Moreover, they present a poisoning risk in young children because of their attractive packaging, taste, and occasional lack of safety caps. Despite these concerns, effective advertising combined with the parental desire to ameliorate symptoms has ensured a continued market for these medications. This review concentrates on the current use of these medications and discusses some of the commonly used medications and their hazards in children.

PREVALENCE

Relatively few studies have described the epidemiology of over-the-counter medication use among children. This is surprising given the prominence and expenditures for over-the-counter medications in the health care system. There are more than 800 OTCs for the common cold (Lowenstein & Parrino, 1987),

The assistance of Andrea Talbot and Wenda MacDonald, PhD, with the manuscript is gratefully recorded.

and over 100 for treatment of diarrhea (Dukes, 1990). Consumers in the United States spend almost $2 billion per year on cough and cold remedies alone (Rosendahl, 1988). It has been estimated that 70% of illnesses are treated with nonprescribed medicines (Knapp & Knapp, 1972). However, most studies have focused on either older adults (Chrischelles et al., 1992; Conn, 1991) or the adult, nonelderly population (Johnson & Pope, 1983; Bush & Rabin, 1976).

Several studies from different countries have indicated high rates of use among children, covering the newborn to 18 years. The initial studies in this area came from Great Britain. Jefferys et al. (1960) found that 66% of the surveyed parents reported giving OTCs to their children in the 4 weeks before the interview. In a later survey of British households, Dunnell and Cartwright (1972) found that 48% of the children had received OTCs in the 2-week recall period before the interview. They also reported that the number of medications increased as the number of symptoms increased.

In the United States, prevalence of use has been examined at both the community and national levels. In a community survey conducted by Haggerty and Roghmann (1972) in Monroe County, New York, it was found that 22% of the children had taken OTCs over a 2-day period, although vitamin use was included. Kovar (1985), using data from the 1981 National Health Interview Survey, reported that about 63% of U.S. children had some form of medication during any 2-week period in 1981. However, the definition of medication use included either prescribed or recommended medications or vitamin–mineral supplements. The prevalence of OTCs was not reported separately. Kogan et al. (1994), reporting on a U.S. national survey of preschool aged children, found that about 54% of children had received OTCs in the 30 days before the interview. In addition, many children in the study by Kogan et al. (1994) had been given more than one OTC in the preceding 30 days. About 40% of the OTC medication users received 2 medications, while 5% of survey respondents reported giving their children 3 or 4 different kinds of OTCs.

Different U.S. national studies have also indicated a high prevalence of use among types of medications. Kovar (1985) found that among children in the age range 3–6 years old, 24% had been given a pain remedy in the previous 2 weeks. Kogan et al. (1994) also found that use of these classes of medications was extensive among 3-year-olds: 42% had been given a pain medication in the 30 days prior to the interview. However, there appeared to be temporal changes in the types of pain medications given to children, probably due to the warnings of the increased risk of Reye's syndrome with the use of aspirin (Maheady, 1989). In a 1984 survey of prescribed and nonprescribed drug use among children in Great Britain, Rylance et al. (1988) found that aspirin accounted for about 15% of all drugs taken in any one week, whereas Kogan et al. (1994), using 1991 data, found that only 1.2% of the children in their survey were given aspirin in the 30 days before the interview. The apparent lack of changes in the prevalence of cough and cold medication use among children is less reassuring. Using 1981 data, Kovar (1985) found 26% of children from 3 to 6 years old were given a cough or cold medication in the 2 weeks prior to interview. Kogan et al. (1994) found 36% of 3-year-olds in the United States had been given a cough or cold medication in the preceding 30 days. Evidence that these medications are inefficacious (Smith & Feldman, 1993; Sakchainanont et al., 1990; Hutton et al., 1991) and may, in some circumstances, have adverse effects (Blake, 1993) has apparently done little to dampen enthusiasm for their use.

Although the use of antidiarrhea medications has been relatively infrequent when examined (Kogan et al., 1994), it is of concern that about 15% of preschool aged children who had symptoms of diarrhea were reportedly given antidiarrhea agents, since their use is not indicated for children under the age of 3 due to the potential side effects (Ginsberg, 1973; Rumack & Temple, 1974; Pickering, 1991).

The results of studies that have examined factors associated with over-the-counter medication use in children have been relatively congruent. Studies by Jefferys et al. (1960), Maiman et al. (1982, 1986), and Kogan et al. (1994) indicated that the use of OTCs increased with higher education and income level. The reasons for these findings are a matter of some speculation. These associations may occur because higher income families tend to have greater resources for purchasing OTCs, or advertising for OTCs may be targeted at these groups. Craig (1992) demonstrated that television advertisement of OTCs took advantage of stereotypical images of women as home medical caregivers, and raised the question of whether female consumers were being encouraged to overuse OTC medications as a way of gaining the family's love and respect.

Kogan et al. (1994) also found a significant association between lack of insurance and OTC medication use after controlling for recent reported child illnesses. This suggests that women without health insurance may be more dependent on OTC medications perhaps as a substitute for treatment in the formal health care system. Viewed in conjunction with the findings that more affluent women tend to give OTCs to their children, the association between health insurance and OTC usage indicates that there may be distinctive subpatterns influencing the use of OTCs. The ubiquity of over-the-counter medications in the treatment of pediatric illnesses combined with the relative paucity of studies on their use in this population suggests that much work still needs to be done on the epidemiology of OTC use.

MEDICATION TYPES

Cough and Cold Medications

A simple upper respiratory infection or "cold" is a common experience in childhood. The average preschool child experiences 6–8 infections a year, with an average of 12 if the child attends nursery school (Dingle et al., 1964). Over 200 viruses can cause the common cold syndrome, with no specific or distinctive clinical characteristics (Smith, 1995). Nasal irritation often begins the syndrome, followed by a vaguely scratchy throat, followed by sneezing and profuse nasal discharge. There is often a low-grade fever accompanying this illness, but occasionally the temperature can be as high as 103–104°C. There is often interference with eating and sleeping, especially in infants. Symptoms usually last from 4 to 7 days. Many parents understandably want to ease these minor but irritating symptoms and therefore turn to OTC medications for help. Antipyretics, analgesics, and combinations of these are used for this common illness. There are over 800 medications to choose from, all having varying sites of action. The breadth of choice is all the more striking given the limited information demonstrating their effectiveness (Smith & Feldman, 1993). The groups of commonly used medications can be conveniently divided into the following groups. Antipyretics are reviewed separately, but it should be noted that they are often combined with many of the cold medications.

Sympathomimetics Sympathomimetic substances produce a physiological response similar to the catecholamines. These substances directly stimulate the alpha- and beta-adrenergic receptors or release endogenous noradrenaline from presynaptic nerve terminals. These medications are found in bronchodilators, stimulants, and appetite suppressants, as well as in cough and cold medications. The main beneficial action in the common cold is the vasoconstriction of the nasal mucosa, which reduces edema and therefore improves nasal airflow, thus lessening the symptoms of the infection. These substances have only been critically examined in combination preparations with antihistamines and analgesics and have been shown to reduce nasal symptoms, aching, cough, and sore throat (Jaffe & Grimshaw, 1983; Weippl, 1984).

Ephedrine and pseudoephedrine have a rapid onset of action, reaching a peak at 1–2 h after ingestion. They can produce hypertension and tachycardia, which are reportedly more often a problem in adults than children. Signs and symptoms of toxicity with these medications include headache, nausea, vomiting, diaphoresis, agitation, psychosis, hypertension, seizures, tremulousness, and rhabdomyolysis (Cetaruk & Aaron, 1994).

Phenylpropanolamine is another sympathomimetic compound similar to ephedrine and pseudoephedrine. This substance is used in combination with analgesics, antihistamines, and anticholinergic and antitussive medications. Its major problem is hypertension, especially in combination with caffeine intake (Brown et al., 1991). Side effects can be seen both in therapeutic doses and in excessive ingestion, particularly central nervous system (CNS) overstimulation and hypertension.

A similar picture is produced by phenylephrine, which is also used in cold medications. It can be effective as a nasal preparation, although its absorption can be unpredictable.

Imidazolines (oxymetazoline, naphazoline, tetrahydrozoline, and xylometazoline) are mainly used as topical agents with primarily a vasoconstrictive action. There are conflicting reports as to their effectiveness (Winther et al., 1983; Akerlund et al., 1989). They are used in sinusitis, allergic rhinitis, colds, and ocular irritations. Generally these medications are not recommended for use in the preschool age group. In occasional use topical vasoconstrictors are promptly effective, although when used continuously (especially over 3–4 days) they can result in rebound nasal congestion and thus prolong the cold. Signs of clinical toxicity cover a wide range of symptoms from lethargy, somnolence, pallor, and cool extremities to miosis, bradycardia, hypotension, loss of consciousness, and respiratory depression (Liebelt & Shannon, 1993).

Cough Suppressants Dextromethorphan is one of the most commonly used antitussive medications. It has a very similar action to codeine, which suppresses the cough center in the medulla oblongata. It continues to be used as codeine for its antitussive action, despite evidence that it is ineffective (Taylor et al., 1993). Dextromethorphan-containing cold/cough preparations are frequently prescribed and bought over the counter for use in children. Although generally considered safe, dextromethorphan has been shown to cause CNS side effects, including hyperexcitability, increased muscle tone, and ataxia. Two deaths have been reported with intentional dextromethorphan overdose (Pender & Parks, 1991). As with any narcotic derivative, there is always the potential for abuse with either codeine or dextromethorphan (Helfer & Oksuk, 1990; Fleming, 1986).

Antihistamines Antihistamines exert their effects via their anticholinergic activity in drying up the respiratory secretions. The anticholinergic effect is often weak given the dosages used in most OTC preparations. The antihistamine effect is also probably minimal given the lack of histamine released during a viral nasopharyngitis (Gaffey et al., 1988). These medications are used for the treatment of allergic conditions, the common cold, motion sickness, and dysmenorrhea. They are also used as sedatives. The most common groups include the ethanolamines (diphenhydramine, dimenhydrinate), the alkylamines (chlorpheniramine, brompheniramine), and the piperidines (terfenadine, astemizole). They have been demonstrated to be most useful when combined with a decongestant and only in older children, adolescents, and adults (Smith & Feldman, 1993). In therapeutic doses, the antihistamines can act as a sedative. Many parents consider this to be a useful effect and wrongly attribute it to a relief of symptoms. Occasionally the use of these medications may result in paradoxical overstimulation, especially in children. In overdose, CNS stimulation predominates but anticholinergic toxicity is also present in varying degrees (Cetaruk & Aaron, 1994).

Expectorants The expectorants purport to reduce secretion viscosity in order to promote more effective expectoration. Guaifenesin is the most widely used of this type of medication. In some formulations this substance is combined with a cough suppressant, which defies logical reasoning. In one clinical trial, young adults with a cold perceived a minor reduction in the quantity of sputum after using this medication (Kuhn et al., 1982). There is no evidence demonstrating its effectiveness in children. In some clinical situations, gastric upset is a side effect.

Gastrointestinal Agents

Vomiting and diarrhea are common symptoms experienced by children. In the majority of situations the cause is due to simple self-limited viral gastroenteritis. Typically the illness begins with mild fever and vomiting, followed quickly by frequent watery diarrhea. Generally the illness lasts 5–7 days, with the vomiting and fever predominating at the start and then changing to diarrhea toward the end. Vomiting and diarrhea can be caused by other infections in children, especially in the younger ages. As a result, there is considered to be no role for the routine use of these medications (Bass, 1996). Despite this recommendation, Morrison and Little (1981) found that 41 (21%) of 181 children admitted to hospital with a diagnosis of acute viral gastroenteritis had been treated with antiemetics before admission. It appears that parents want to control and limit these symptoms and are aware of the risks of dehydration. Families use antiemetics or antidiarrheals for this illness. Many products are available for their use.

Antiemetics   Dimenhydrinate is a commonly used antiemetic. As an H_1 antagonist, its principal effect is to counteract the effects of histamine, including smooth muscle contraction. All of the H_1 antagonists bind to the H_1 receptors in the central nervous system and may be responsible for depression or stimulation. The mechanism of its antiemetic action is unclear, although it is thought to have a depressant effect on labyrinthine function. Dimenhydrinate is indicated for motion sickness, radiation sickness, postoperative vomiting, drug-induced vomiting, and Meniere's disease, but not viral gastroenteritis. It is most effective prior to the development of nausea and vomiting. At one point it was used for colic, but this practice has been discouraged because of a possible link with apparent life-threatening events (Hardoin et al., 1991). It is not recommended for acute vomiting, yet many families have it available and use this medication. However, in some situations, the cause of the vomiting may be more serious and may result in considerable harm if appropriate treatment is delayed. Anquist and colleagues (1991) reviewed 148 children who presented to a children's hospital emergency room with vomiting. Twenty-one (14%) of these children had received dimenhydrinate and were more likely to present 12 h later than those who did not take the medication ($p < .01$). The range of diseases treated with this medication included asthma, pelvic inflammatory disease, and urinary tract infection. Anquist et al. (1991) concluded that use of dimenhydrinate is associated with a risk of delay in the diagnosis of a treatable medical condition. In overdose, dimenhydrinate will result in drowsiness and atropine-like effects, although in some cases CNS stimulation is seen.

Antidiarrheals   Acute diarrhea continues to be a common symptom in children. In developed countries, it is likely to be mild, although it occasionally will result in significant dehydration, especially in young infants. The usual symptoms include a sudden onset of vomiting followed by frequent, watery stools. In acute gastroenteritis, the usual cause is an enteric viral pathogen. In children other diseases such as ear or urinary tract infections may mimic this syndrome. Therefore, the risk of inappropriate or delayed treatment is increased.

The principal treatment of acute diarrhea is oral fluid therapy followed by appropriate early refeeding with cereal solutions. In general., there are few indications for antidiarrheal agents. Despite this, there are over 100 OTC antidiarrheals available for children, and they are purchased principally for the treatment of acute infectious diarrhea.

Loperamide is the most commonly used antidiarrheal in the world. Its action is to reduce propulsive peristalsis and increase intestinal transit time by binding to the opiate receptor in the gut wall. This medication has been progressively switched to nonprescription status in an increasing number of countries (Fletcher et al., 1995). It has shown to be effective in acute nonspecific diarrhea, acute functional diarrhea, and traveller's diarrhea. The main serious side effects with this medication occur in young children. The main problems include those related to CNS depression including drowsiness, lethargy, convulsions, and respiratory depression (Bhutta & Tahir, 1990; Harries & Rossiter, 1969; Rosenton et al., 1973). In one recent trial loperamide was shown to reduce the severity and duration of diarrhea in infants with infectious diarrhea but produced a 20% incidence of significant and potentially serious side effects such as paralytic ileus and persistent drowsiness (Motala et al., 1990). Currently it is not recommended for children with acute diarrhea.

Kaolin and pectin are traditional treatments for diarrhea that utilize the absorbent qualities of these substances. Kaolin is a clay, and pectin is derived from the rind of apples and citrus fruit (Dupont, 1985). It has been demonstrated that these substances increase the form and content of the stools in acute infantile diarrhea (Portnoy et al., 1976). It has not been proven whether passing larger stools with more form is beneficial to the resolution of the diarrhea. No specific side effects have been reported.

Bismuth subsalicylate has long been known to be effective in the management of traveller's diarrhea (Gorbach, 1990). Its mechanism of action is unknown, but some explanations include preventing the attachment of the microbes to the intestinal surface, direct antimicrobial effects, toxin inactivation, antinflammatory effects, or binding of the bile acids that may contribute to the diarrhea (Gryboski et al., 1985). Only recently has it been shown to be effective in infants and children (Soriano-Brucher et al., 1991; Figueroa-Quintanilla et al., 1993). Compared with many other OTC medications, it is effective in reducing stool frequency, need for treatment, and duration of acute diarrheal disease. To be effective it needs to be given every 4 h, which may impair compliance. No reports of toxicity have been reported in children, but there may be a potential problem of bismuth accumulation or salicylate toxicity (discussed later) if given frequently (Gorbach, 1990).

Antipyretics and Analgesics

Fever is a common response to a variety of pediatric infections, and most families treat their child immediately with antipyretic therapy. Fever engenders considerable anxiety in parents, leading them to seek medical attention early and frequently (Schmitt, 1980; Kramer et al., 1985). Many families aggressively treat even a mild elevation of temperature (37.0–37.8°C) with antipyretics—and this range includes the normal range (Schmitt, 1980). There has been considerable debate whether pharmacological reduction of fever is beneficial or hazardous to the host (Drwal-Klein & Phelps, 1992). Fever is a useful sign of illness, and reduction may mask the onset of a serious disease. High fever may be a useful mechanism for fighting bacterial infections. Conversely, an elevated temperature is uncomfortable and adverse events may be more likely, such as seizures, confusion, or dehydration. The common OTC medications used for this symptom are acetaminophen and aspirin.

Acetaminophen   Acetaminophen is a weak inhibitor of prostanoid biosynthesis with antipyretic and analgesic properties. It is rapidly absorbed from the gastrointestinal tract, reaching peak plasma concentrations in 60 min. Because of the association between aspirin and Reye's syndrome, acetaminophen has virtually replaced all OTC analgesic preparations. Acetaminophen is found in a variety of forms on its own but also combined with a variety of substances in cough and cold preparations. It is safe in therapeutic doses, although there are isolated problems in a few patients resulting bone marrow depression or allergic phenomenon (Kasco & Terezhalmy, 1994). Hepatotoxicity can occur in chronic overdose just slightly above therapeutic levels. In acute overdose, its major problem is hepatotoxicity, but the incidence of serious sequelae from this is much less in children than in adults (Drwal-Klein & Phelps, 1992).

Acetylsalicylic Acid (ASA or Aspirin)   Prior to the link with Reye's syndrome, ASA was a common analgesic and antipyretic used in children. As with acetaminophen, it is still occasionally used for the usual viral illnesses but its major role is in the treatment of more serious illnesses such as Kawasaki disease or juvenile rheumatoid arthritis. Aspirin is useful in treating the dull, throbbing pain of inflammation produced by prostanoids, which sensitizes the peripheral nerve endings to endogenous pain-producing substances (Kasco & Terezhalmy, 1994). As an inhibitor of prostanoid synthesis it reduces the perception of this pain. It also has a very useful antipyretic effect, mainly by altering the hypothalamic heat-regulating center. Its anti-inflammatory effects are chiefly due to the reduction of vasodilatation, thereby reducing vascular permeability. Aspirin is rapidly absorbed from the stomach and the upper small intestine, reaching peak values 2 h later. The adverse effects associated with aspirin have generally overshadowed the therapeutic benefits. Perhaps the most serious and fortunately rarest association is Reye's syndrome. This is an acute encephalopathy in association with fatty degeneration of the liver. It is not clear whether this disease is caused or worsened by aspirin. In Europe the disease affects babies and preschool children, whereas in North America school age children are affected with influenza-like illnesses (Glasgow & Moore, 1993). The incidence has declined since public warnings have been issued about the use of aspirin in acute febrile illnesses (Glasgow & Moore, 1993). Another problem associated with aspirin use is the development of hypersensitivity in those with preexisting allergic intolerance, such as asthma, nasal polyps, urticaria, and multiple allergies. The usual history is urticaria, angioedema, and wheezing or severe rhinorrhea occurring within 3 h of aspirin ingestion (Kasco & Terezhalmy, 1994). Gastric intolerance is usually a problem with adult populations. In overdose the duration of poisoning is critical in determining the outcome. A short, acute poisoning is characterized by pH disturbances and mild symptoms and the intoxication is tolerated (Yip et al., 1994). In contrast, the chronic overdosing of a child can present a more ominous scenario and is responsible for increasing morbidity. Usually, the medication is given at the onset of a febrile illness or other minor coincidental illness and several days may elapse before the child is brought to medical attention. At that time the child appears very ill and the diagnosis is often thought to be overwhelming sepsis because of the symptoms of CNS disturbance, tachypnea, fever, and leukocytosis (Yip et al., 1994). However, the presence of acidosis and severe hypokalemia in combination with these usually points to the diagnosis of chronic salicylism.

CONCLUSIONS

OTC medications are commonly used for the treatment of the everyday discomforts of childhood illnesses. Home treatment does not conform to the usual biomedical model that exists in hospitals, medical textbooks, and scientific journals. Many other factors contribute to the choice of treatment parents select for their children. The powerful effect of the media is to perpetuate the myth that medication must be bought for the treatment of every symptom. Parents are very susceptible to this advertising because of their desire to do something (Hutton et al., 1991; Gadomski, 1994). Frequently this produces demands on health care providers, who are coerced into recommendations or providing prescriptions (Gadomski & Rubin, 1993). In addition, many families mistakenly believe that certain OTC medications "cure" the viruses that cause an infection such as the common cold (Flannery, 1981). Part of this may be a belief system in "scientific" OTC medications, which have replaced the folk cures of previous generations (Helman, 1978). Finally, there is a belief that many of these medications have been rigorously evaluated and actually work. With a few exceptions, these medications are not recommended and may actually worsen the illness in the child by development of side effects, delaying the diagnosis, or causing further problems with the disease process. Physicians and pharmacists need to take the initiative to inquire about these medications in their patients in order to understand why they are used. Misconceptions need to be corrected and clear advice provided on the rational use of OTC medications. Further areas of research will need to address these issues and include the influence of health care provider recommendations on parental administration of OTCs to their children and the effect of illness severity and duration on use of OTCs.

REFERENCES

Akerlund, A., Klint, T., Olen, L., and Rundcrantz, H. 1989. Nasal decongestant effect of oxymeta-zoline in the common cold: An objective dose-response study in 106 patients. *J. Laryngol. Otol.* 103:743–746.

Anquist, W., Panchanathan, S., Rowe, P. C., Peterson, R. G., and Sirnick, A. 1991. Diagnostic delay after dimenhydrinate use in vomiting children. *Can. Med. Assoc. J.* 145:965–968.

Bass, D. M. 1996. Rotavirus and other agents of viral gastroenteritis. In *Nelson textbook of pediatrics*, eds. R. E. Behrman, R. M. Kliegman, and A. M. Arvin, pp. 914–916. Philadelphia: W. B. Saunders.

Bhutta, T. I., and Tahir, K. I. 1990. Loperamide poisoning in children. (letter). *Lancet* 335:363.

Blake, K. D. 1993. Dangers of the common cold treatments in children. *Lancet* 341–640.

Brown, N. J., Ryder, D., and Branch, R. A. 1991. A pharmacodynamic interaction between caffeine and phenylpropanolamine. *Clin. Pharmacol. Ther.* 50:363–371.

Bush, P. J., and Rabin, D. L. 1976. Who's using nonprescribed medicines? *Med. Care* 14:1014–1023.

Cetaruk, E. W., and Aaron, C. K. 1994. Hazards of nonprescription medications. *Emerg. Med. Clin. North Am.* 12:483–510.

Chrischelles, E. A., Foley, D. J., Wallace, R. B., Lemke, J. H., Semla, T. P., Hanlon, J. T., Glynn, R. J., Ostfeld, A. M., and Guralnik, J. M. 1992. Use of medications by persons 65 and over: Data from the established populations for epidemiologic studies of the elderly. *J. Gerontol.* 47:M137–M144.

Conn, V. S. 1991. Older adults: Factors that predict the use of over-the-counter medication. *J. Adv. Nurs.* 16:1190–1196.

Craig, R. S. 1992. Women as home caregivers: Gender portrayal in OTC drug commercials. *J. Drug Educ.* 22:303–312.

Dingle, J. H., Badger, G. F., and Jordan, W. S. 1964. *Illness in the home: A study of 25,000 illnesses in a group of Cleveland families.* Cleveland, OH: Press of Western Reserve University.

Drwal-Klein, L. A., and Phelps, S. J. 1992. Antipyretic therapy in the febrile child. *Clin. Pharm.* 11:1005–1021.

Dukes, G. E. 1990. Over-the-counter antidiarrheal medications used for the self-treatment of acute nonspecific diarrhea. *Am. J. Med.* 88:24S–26S.

Dunnell, K., and Cartwright, A. 1972. *Medicine takers, prescribers and hoarders.* New York: Routledge & Kegan Paul.

Dupont, H. L. 1985. Nonfluid therapy and selected chemoprophylaxis of acute diarrhea. *Am. J. Med.* 78:81–90.

Figueroa-Quintanilla, D., Salazar-Lindo, E., Sack, R. B., Leon-Barua, R., Sarabia-Arce, S., Campos-Sanchez, M., and Eyzaguirre-Maccan, E. 1993. A controlled trial of bismuth sub-salicylate in infants with acute watery diarrheal disease. *N. Engl. J. Med.* 328:1653–1658.

Flannery, D. B. 1981. Parents' knowledge about acetaminophen. *J. Pediatr.* 98:851.

Fleming, P. M. 1986. Dependence on dextromethorphan hydrobromide. *Br. Med. J.* 293:597.

Fletcher, P., Steffen, R., and DuPont, H. 1995. Benefit/risk considerations with respect to OTC-descheduling of loperamide. *Arzneim-Forsch. Drug Res.* 45:608–613.

Gadomski A. 1994. Rational use of over-the-counter medications in young children. *J. Am. Med. Assoc.* 272:1063–1064.

Gadomski, A. M., and Rubin, J. D. 1993. Cough and cold medicine use in young children: A survey of Maryland pediatricians. *Maryland Med. J.* 42:647–650.

Gaffey, M. J., Kaiser, D. L., and Hayden, F. G. 1988. Ineffectiveness of oral terfenadine in natural colds: Evidence against histamine as a mediator of common cold symptoms. *Pediatr. Infect. Dis.* 7:223–228.

Ginsburg, C. M. 1973. Lomotil intoxication. *Am. J. Dis. Child.* 125:241–242.

Glasgow, J. F. T., and Moore, R. 1993. Reye's syndrome 30 years on. *Br. Med. J.* 307:950–951.

Gorbach, S. L. 1990. Bismuth therapy in gastrointestinal diseases. *Gastoenterology* 99:863–875.

Gryboski, J. D., Hillemeier, A. C., Grill, B., and Kocoshis, S. 1985. Bismuth subsalicylate in the treatment of chronic diarrhea of childhood. *Am. J. Gastroenterol.* 80:871–876.

Haggerty, R. J., and Roghmann, K. J. 1972. Noncompliance and self-medication: Two neglected aspects of pediatric pharmacology. *Pediatr. Clin. North Am.* 19:101–115.

Hardoin, R. A., Henslee, J. A., Christenson, C. P., Christenson, P. J., and White, M. 1991. Colic medication and apparent life-threatening events. *Clin. Pediatr.* 30:281–285.

Harries, J. T., and Rossiter, M. 1969. Fatal Lomotil poisoning (letter). *Lancet* 1:150.

Helfer, J., and Oksuk, M. K. 1990. Psychoactive abuse potential of Robitussin-DM. *Am. J. Psychiatry* 147:672–673.

Helman, C. G. 1978. "Feed a cold, starve a fever"—Folk models of infection in an English suburban community, and their relation to medical treatment. *Cult. Med. Psychiatry* 2:107–137.

Hutton, N., Wilson, M. H., Mellits, E. D., Baumgardner, R., Wissow, L. S., Bonuccelli, C., Holtzman, N. A., and DeAngelis, C. 1991. Effectiveness of an antihistamine-decongestant combination for young children with the common cold: A randomized controlled clinical trial. *J. Pediatr.* 118:125–130.

Jaffe, G., and Grimshaw, J. J. 1983. Randomized single-blind trial in general practice comparing the efficacy and palatability of two cough linctus preparations, "Pholcolix" and "Actifed" compound, in children with acute cough. *Curr. Med. Res. Opin.* 8:594–599.

Jefferys, M., Brotherston, J. H. F., Cartwright, A. 1960. Consumption of medicines on a working-class housing estate. *Br. J. Prev. Soc. Med.* 14:64–76.

Johnson, R. E., and Pope, C. R. 1983. Health status and social factors in nonprescribed drug use. *Med. Care* 21:225–233.

Kasco, G., and Terezhalmy, G. T. 1994. Acetylsalicylic acid and acetaminophen. *Dent. Clin. North Am.* 38:633–644.

Knapp, D. A., and Knapp, D. E. 1972. Decision-making and self-medication: Preliminary findings. *Am. J. Hosp. Pharm.* 29:1004–1012.

Kogan, M. D., Pappas, G., Yu, S. M., Kotelchuck, M. 1994. Over-the-counter medication use among preschool aged children in the United States. *J. Am. Med. Assoc.* 272:1025–1030.

Kovar, M. G. 1985. Use of medications and vitamin-mineral supplements by children and youths. *Public Health Rep.* 100:470–473.

Kramer, M. S., Naimark, L., and LeDuc, D. G. 1985. Parental fever phobia and its correlates. *Pediatrics* 75:1110–1113.

Kuhn, J. J., Hendley, J. O., Adams, K. F., Clark, J. W., and Gwaltney, J. M. 1982. Antitussive effect of guaifenesin in young adults with natural colds. *Chest* 82:713–718.

Liebelt, E. L., and Shannon, M. 1993. Small doses, big problems: A selected review of highly toxic common medications. *Pediatr. Emerg. Care* 9:292–297.

Lowenstein, S. R., and Parrino, T. A. 1987. Management of the common cold. *Adv. Intern. Med.* 32:207–234.

Maheady, D. C. 1989. Reye's syndrome: Review and update. *J. Pediatr. Health Care* 3:246–250.

Maiman, L. A., Becker, M. H., Cummings, K. M., Drachman, R. H., and O'Connor, P. A. 1982. Effects of sociodemographic and attitudinal factors on mother-initiated medication behavior for children. *Public Health Rep.* 97:140–149.

Maiman, L. A., Becker, M. H., and Katlic, A. W. 1986. Correlates of mothers' use of medications for their children. *Soc. Sci. Med.* 22:41–51.

Morrison, P. S., and Little, T. M. 1981. How is gastroenteritis treated? *Br. Med. J.* 283:1300.

Motala, C., Hill, I. D., Mann, M. D., and Bowie, M. D. 1990. Effect of loperamide on stool output and duration of acute infectious diarrhea in infants. *J. Pediatr.* 117:467–471.

Pender, E. S., and Parks, B. R. 1991. Toxicity with dextromethorphan-containing preparations: A literature review and report of two additional cases. *Pedatr. Emerg. Care* 7:163–165.

Pickering, L. K. 1991. Therapy for acute infectious diarrhea in children. *J. Pediatr.* 118(suppl. 4):S118–S128.

Portnoy, B. L., Dupont, H. L., Pruit, D., Abdo, J. A., and Rodriguez, J. T. 1976. Antidiarrheal agents in the treatment of acute diarrhea in children. *J. Am. Med. Assoc.* 236:844–846.

Rosendahl, I. 1988. Expense of physician care spurs OTC, self-care market. *Drug Topics* 132:62–63.

Rosenton, G., Freeman, M., Standard, A. L., and Weston, N. 1973. Warning: Use of Lomotil in children. *Pediatrics* 51:132–134.

Rumack, B. H., and Temple, A. R. 1974. Lomotil poisoning. *Pediatrics* 53:495–500.

Rylance, G. W., Woods, C. G., Cullen, R. E., Rylance, M. E. 1988. The use of drugs by children. *Br. Med. J.* 297:445–447.

Sakchainanont, B., Chantarojanasiri, T., and Runagkanchanasetr, S. 1990. Effectiveness of anti-histamines in common cold. *J. Med. Assoc. Thail.* 73:96–100.

Schmitt, B. D. 1980. Fever phobia. *Am. J. Dis. Child.* 134:176–181.

Smith, M. B. H. 1995. Transmission and therapy of common respiratory viruses. *Curr. Opin. Infect. Dis.* 8:209–212.

Smith, M. B. H., and Feldman, W. 1993. Over-the-counter cold medications: A critical review of clinical trials between 1950 and 1991. *J. Am. Med. Assoc.* 269:2258–2263.

Soriano-Brucher, H., Avendano, P., O'Ryan, M., Braun, S. D., Manhart, M. D., Balm, T. K., and Soriano, H. A. 1991. Bismuth subsalicylate in the treatment of acute diarrhea in children: A clinical study. *Pediatrics* 87:18–27.

Taylor, J. A., Novack, A. H., Almquist, J. R., and Rogers, J. E. 1993. Efficacy of cough suppressants in children. *J. Pediatr.* 122:799–802.

Weippl, G. 1984. Therapeutic approaches to the common cold in children. *Clin. Ther.* 6:475–482.

Winther, B., Brofeldt, S., Borum, P., Pedersen, M., and Mygind, N. 1983. Lack of effect on nasal discharge from a vasoconstrictor spray in the common cold. *Eur. J. Respir. Dis.* 64S:447–448.

Yip, L., Dart, R. C., and Gabow, P. A. 1994. Concepts and controversies in salicylate toxicity. *Emerg. Med. Clin. North Am.* 12:351–364.

Index